BABIES IN BOTTLES

BABIES IN BOTTLES

Twentieth-Century Visions of Reproductive Technology

Susan Merrill Squier

Rutgers University Press
New Brunswick, New Jersey

Library of Congress Cataloging-in-Publication Data
Squier, Susan M. (Susan Merrill)
    Babies in bottles : twentieth-century visions
  of reproductive technology / Susan Merrill Squier.
      p. cm.
    Includes bibliographical references (p.   ) and index.
    ISBN 0-8135-2116-5 (cloth) — ISBN 0-8135-2117-3 (pbk.)
    1. Human reproductive technology—Social aspects.
    2. Human reproductive technology in literature.   I. Title.
  RG133.5.S98   1994
  306.4'61—dc20
                                                94-10154
                                                    CIP

British Cataloging-in-Publication information available

Illustration 1 reproduced courtesy of Tony Bela, *The Sunday Mail*, Brisbane,
    Australia.
Illustration 2 copyright © 1989 by the New York Times Company. Reproduced by
    permission.
Illustration 4 reproduced by permission of The Putnam Publishing Group from
    Mutation by Robin Cook. Copyright © 1989 by Robin Cook.
Illustration 9 reproduced from *Test Tube Conception* by Carl Moore and Ann
    Westmore (London: George Allen & Unwin, 1984). Copyright © 1983 by Carl
    Moore and Ann Westmore.

nwst
IA6M4768

To Gowen, Caitlin, and Toby—
Intrepid Travelers—
and to Howard,
connoisseur of slime molds,
bats, and the *Wall Street Journal*

Contents

Illustrations

Preface and Acknowledgments

Researching and writing *Babies in Bottles* took five years, and extended to three continents. It is with great pleasure that I acknowledge some of the people and institutions who helped along the way. Although origins are always mysterious, and intellectual origins no less so than biological ones, two collaborative experiences were important germinal moments for this book. The experience of working with Helen Cooper and Adrienne Munich on our co-edited collection *Arms and the Woman: War, Gender, and Literary Representation* (Chapel Hill: University of North Carolina Press, 1989) first led me to think about the modern and postmodern disjunction between sexuality and reproduction. Then Lou Charnon-Deutsch's translation of Laura Freixas extraordinary intrauterine parable "Mi mamá me mima" ("My Mama Spoils Me") sensitized me to the cultural prominence of fetal images, catalyzing my essay "Fetal Voices: Speaking for the Margins Within," an early foray into an issue whose modernist precursor I explore in this book. Angela Ingram and Daphne Patai were early supporters of my work on Charlotte Haldane, for which I thank them. I thank Jane Marcus for giving me the opportunity to present work in progress with Barbara Katz Rothman at the Forum of Technologies of Birth held at the Graduate Center of the City University of New York, and I thank Elaine Hoffman Baruch and Rayna Rapp for their generous responses to other work in progress. Finally, the Humanities Institute at Stony Brook provided crucial intellectual stimulation and good fellowship. In particular, I want to thank members of the faculty seminar "Motherhood and Representation" led by E. Ann Kaplan; participants in the Humanities Institute Conference "Reproductive

Technologies: Narrative, Gender, Culture"; and members of Jennifer Terry's "Bodies" seminar, especially Ira Livingston.

For the administrative leave that made my research year in London possible, I thank J. R. Schubel, Don Ihde, and David Sheehan. For research help and other support in London, I wish to thank Karl Miller of University College, London; Lesley A. Hall, of the Wellcome Institute for the History of Medicine; Dennis Doughan of the Fawcett Library; and the staffs of the D.M.S. Watson Library of University College, London, the Wellcome Institute for the History of Medicine Library, and the Manuscript Room of the National Library of Scotland. Robert Young, of Free Associations Books, was a splendid guide to the new territory of science studies. Anna Werrin helped me understand, and gain access to, the records of the Human Fertilisation and Embryology Debate in Parliament, and I thank her for her insider's perspective on the House of Commons. I thank Jacqueline Rose and Rachel Bowlby for the opportunity to present part of Chapter 1 at the Cultural Studies Program of the University of Sussex on 6 June 1990, and Miriam Diaz-Diocaretz and the organizers of the International Conference on Women, Culture, and the Arts in Dubrovnik on 16–20 April 1990, for the opportunity to present part of Chapter 3. For social as well as intellectual sustenance in London I wish to thank Marina Benjamin, Madeline and Ian Crispin, Lesley Hall, Maggie Humm, Bonnie Kime Scott, and Eve Stoddard. For helping me track down many of the rare texts this book required, I thank booksellers Elizabeth Crawford, of London, and Ian Patterson, of Cambridge, England. Finally, I thank Naomi Mitchison for her generous support of my project and for her inspiring example of a complex, scientifically informed, and scientifically critical feminism.

A Fulbright Senior Research Scholar fellowship made possible my research year in Melbourne, Australia, and I am grateful to the Fulbright organization for the chance to consider feminist issues from such a valuable new vista. I want to thank Maila Stivens, director of Women's Studies at the University of Melbourne, and Stuart MacIntyre, of the History Department at the University of Melbourne, for providing such a hospitable environment for research and writing. And I want to thank Gaye Balwin, Peter Fitzpatrick, Pat Grimshaw, Alec Hyslop, Joel Kahn, Diane Kirkby, Maila Stivens, and Chips and Aude Sowerwine for making Melbourne home. For a variety of kindnesses I also want to thank Rebecca Albury, Purisottima Billimoria, Dipesh Chakrabarty, Heather Dietrich, Sandy Gifford, Sneja Gunew, Beryl Langer, Pauline Nestor, Susie Nixon, Philipa Rothfield, Robyn

Rowland, and Lisa Woll. I am grateful to the members of the Feminism and Legal Studies seminar for providing a stimulating site for feminist debate and communion. Ann Brody kept me informed about American news coverage on RT issues while I was in Melbourne, and I thank her. My thanks also to the following institutions for giving me the opportunity to give seminars on work in progress: the History and English Departments of the University of Melbourne; the Sociology/ Anthropology Department of Monash University; the Department of English of LaTrobe University; the Department of Literature, Deakin University; the Department of English of the University of Wollongong (New South Wales); the Department of English of the University of Canterbury, New Zealand; and the Department of Women's Studies, University of Otago, New Zealand; as well as the Melbourne Feminist History Group and the Second Annual Australian Women's Studies Convention. The opportunity to revisit Australia in 1992, as Distinguished Visiting Fellow at LaTrobe University, gave me the chance to think through some of the postmodern implications of modernist reproductive technologies, and I want to thank Lucy Frost, Kay Torney, and John Wiltshire of the Department of English, LaTrobe University, for nominating me for that fellowship and inviting me to present my work at the Postmodern Body Conference at LaTrobe. Leslie Mitchner has been the ideal editor: incisive, enthusiastic, supportive. I thank the following people for helping in ways each knows best: Syril Blechman, Judy Cartwright, Joan Korins, Dusa McDuff, and Lynn Owen. Finally, last, first, and always, I thank Gowen.

The following early versions of parts of chapters from this book have appeared in print, and I thank the following editors and publishers for permission to reprint revised versions of them:

"Conceiving Difference: Reproductive Technology and the Construction of Identity in Two Contemporary Fictions," *A Question of Identity: Women, Science, and Literature,* ed. Marina Benjamin (New Brunswick, N.J.: Rutgers University Press, 1993), 97–118.

"The [Impregnable] Mother of All Battles: War, Reproduction, and Visualisation Technology," *Meridian* 12 (May 1993): 3–9.

"Representing the Reproductive Body," *Meridian* 12 (May 1993): 29–45.

"Sexual Biopolitics in *Man's World*: The Writings of Charlotte Haldane," *Rediscovering Forgotten Radicals: British Women Writers 1889–1939,* ed. Angela Ingram and Daphne Patai (Chapel Hill: University of North Carolina Press, 1993), 137–155.

BABIES IN BOTTLES

Introduction

REPRODUCTIVE TECHNOLOGY
AND REPRESENTATION

> Any image that is important to a culture constitutes an arena of
> ideological construction rather than simple consolidation.
> —MARY POOVEY, *Uneven Developments*

BABIES IN BOTTLES

While I was researching and writing this book, I collected images of
babies in bottles. My collection now extends from the photographic to
the pictorial to the textual. A sociologist in Wollongong, Australia, sent
me the fanciful photograph of a baby in a bottle that graced the cover
of the January 1986 issue of the American magazine *New Age*.[1] In
1989, I found a woodcut version of a similar image in the *New York
Times Book Review*, accompanying Stephen Dobyns's review of Katherine Dunn's novel *Geek Love*.[2] And in 1991, a woman I met at a conference on surrogacy, in Melbourne, Australia, sent me the schematic
drawing of extrauterine gestation or ectogenesis that appeared in the
Sunday Mail.[3] (Ill. 1–3.) Finally, as an example of a textual image, there
is this passage from Robin Cook's thriller, *Mutation*:

> On a long bench built of rough-hewn lumber sat four fifty-gallon glass
> tanks. The sides were fused with silicone. The tanks were illuminated
> by heat lamps and gave off the eerie blue light as it refracted through
> the contained fluid. . . . Inside each one and enveloped in transparent
> membranes were four fetuses, each perhaps eight months old, who
> were swimming about in their artificial wombs. . . . In each tank the
> placentas were plastered onto a plexiglass grid against a membrane bag
> connected to a sort of heart-lung machine. Each machine had its own
> computer, which was in turn attached to a protein synthesizer.[4]

1

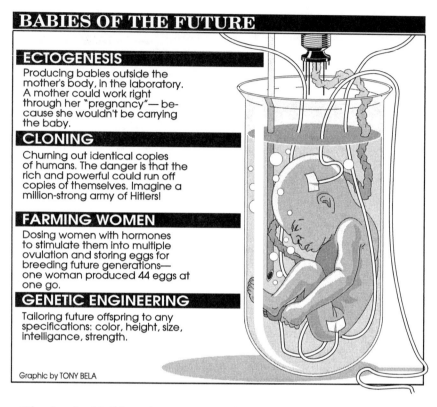

BABIES OF THE FUTURE

ECTOGENESIS
Producing babies outside the mother's body, in the laboratory. A mother could work right through her "pregnancy"—because she wouldn't be carrying the baby.

CLONING
Churning out identical copies of humans. The danger is that the rich and powerful could run off copies of themselves. Imagine a million-strong army of Hitlers!

FARMING WOMEN
Dosing women with hormones to stimulate them into multiple ovulation and storing eggs for breeding future generations—one woman produced 44 eggs at one go.

GENETIC ENGINEERING
Tailoring future offspring to any specifications: color, height, size, intelligence, strength.

Graphic by TONY BELA

1 The *Sunday Mail* Imagines Ectogenesis

These images of babies in bottles—just a small sample of the contemporary traffic in such representations—provide a framework for the arguments I make in this book. They all invoke, and literalize, the test-tube baby, the icon of reproductive technology (RT).[5] This set of medical techniques for technologically mediated reproduction includes artificial insemination, endocrine treatment, egg and sperm donation, egg and embryo culture, sperm and embryo freezing, GIFT (gamete intrafallopian transfer), ZIFT (zygote intrafallopian transfer), and genetic engineering, as well as IVF (in vitro fertilization), and it extends imaginatively to the hypothetical techniques of parthenogenesis and cloning.[6] If we spend a little time teasing out the implications of these images of babies in bottles, we can see that they all enact the fantasy of the womb as a see-through container for the previously in-

2 A Fanciful Baby in a Bottle, from *New Age*

visible fetus, but they differ in the meanings they attach to it, whether explicitly or implicitly.

Although both the visual images from the *Sunday Mail* and *New Age* show a baby in roughly the same position, facing left, knees flexed and back curved, the contrast between the illustrations reveals that both "baby" and "bottle" are terms with variable meanings. In its image of

"Babies of the Future," the *Sunday Mail* graphic offers an ambivalent taxonomy of so-called reproductive advances from ectogenesis to genetic engineering. The list of advantages and disadvantages mingles the cheerful functionalism of the ladies' magazines ("A mother could work right through her 'pregnancy' because she wouldn't be carrying the baby") with the horrified hyperbole of Ripley's *Believe It or Not!* ("One woman produced 44 eggs at one go") and the mass-market, class-stratified alarmism of the tabloids ("The danger is that the rich and powerful could run off copies of themselves. Imagine a million-strong army of Hitlers!"). The *Sunday Mail* image clearly figures a fetus, eyes tightly shut, an expression of concentration on its old-man face. But babies of the future are preeminently scientific creations for the *Sunday Mail*, and thus the graphic emphasizes the medical/technical interventions necessary to their production. Electrodes are prominently placed on the fetal hip and head, what appears to be a temperature probe extends down along the left margin of the beaker, and tubing snakes up from the beaker to various support and monitoring instruments beyond the frame of the illustration.

No such scientific instrumentation mars the serenity of the cover photograph from *New Age*, which offers a far more antiseptic image of extrauterine gestation. Indeed, it hardly seems to be gestation at all: we have what looks like a three-month-old baby, old enough to have plumped out and lost that wizened, aged look. The baby's eyes are open (another indicator of relative maturity), and its face holds an expression of uncertainty, or is it dubiousness? It lies curved on its back, molding to the sides of the bottle, or is it a test tube, or is it merely a drop of water on the end of a pipette?

The magazine's title and cover copy communicate the new age concern with self-actualization ("achievement-commitment-creative living" proclaims the motto) as well as what Andrew Ross has identified as the new age "love-hate affair with rationalist science" (in the antiseptic, yet vaguely scientific, photograph).[7] Bold print to the left of the cover photo indicates the lead article, whose title and subtitle reveal the journal's skepticism about the technological fix applied to reproduction: "Making Babies: High-tech reproduction offers hope to infertile couples. But at what cost?" Two other articles receiving cover billing suggest the contradictory parameters within which that question might be answered by *New Age* readers: "Breaking in the New Year: 44 Painless Resolutions" and "The Homeless: Local Solutions for a National Problem." As this cover copy suggests, the journal's imperatives to individual self-actualization clash with its commitment to social responsibility, postmodern situated knowledges, and painless progress.[8] In a "complex fusion of the alternative and the middlebrow" that for Ross characterizes new age thinking, *New Age* advertises scientifically mediated personal growth while registering anxiety at its social costs (*Strange Weather*, 27). In contrast to the high-tech cyborg of the *Sunday Mail*, we might characterize the *New Age* image as the human potential pregnancy: invisible, scientifically mediated reproduction with little or no pain, and gain for all.

The healthy pre- and postnatal normalcy of the first two images contrasts dramatically with a third image of babies in bottles: the illustration to the review of *Geek Love*. The *New York Times* artist seems to have read the book in question, for the illustration dramatically embodies the novel's central theme: the deliberate creation of deformed fetuses for exhibit to thrill-seeking audiences. As reviewer Stephen Dobyns explains, "The book centers around . . . a carnival run by Aloysius Binewski and his wife, Crystal Lil, whose attractions include their own children, born as freaks owing to their parents' careful experimentation 'with illicit and prescription drugs, insecticides, and eventually radioisotopes.'"[9] These babies in bottles are strikingly different from the previous two images. The illustration pictures not a "bottle" but a laboratory jar, complete with screw-on lid and gummed label. The jar probably contains not water or amniotic fluid, but formaldehyde. The "babies" it holds are certainly dead. The drapery on the right in the illustration further suggests that these fetuses are being exhibited in a freak show, which enhances their monstrousness. On the chubby shoulders of the classically froggy fetal form are perched two heads, each with that closed-eye look of sleeping concentration, each with the sparse hair extending back from the dome-like brow. When "read" in relation to the novel whose review it illustrates, then, this image contrasts with the futurological anxieties of the first image from the *Sunday Mail*, for it seems rather to speak of the past, testifying to the sins of the parents who produced such monstrous babies.

The deliberate deformation of a fetus in utero is also the subject of the last contemporary image of babies in bottles that I want to consider here. This image is textual, appearing in Robin Cook's novel, *Mutation*. Cook's engineered fetal deformity has a significance quite different from that of the freaks engendered in *Geek Love*, because it is the consequence not of abnormal social behavior but of normal science. The parents in *Geek Love* used medical and environmental poisons to produce their monsters; Cook's hero uses the latest in genetic engineer-

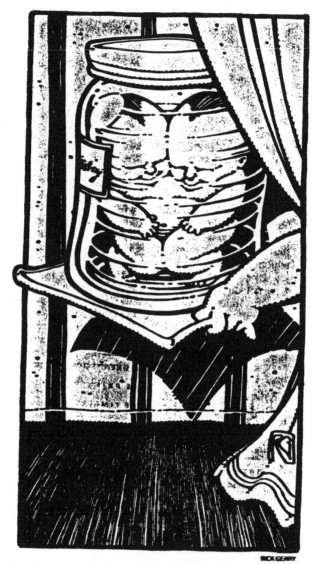

3 A Macabre Version: Deformed Babies in a Specimen Bottle

ing. As the epigraph indicates, *Mutation* was written in homage (albeit exploitative) to Mary Shelley: "How dare you sport thus with life!"[10] Cribbing from Shelley's story of a scientist who usurps female procreative power and creates a monster, Robin Cook gives us Dr. Victor Frank, a physician-researcher with obstetric training and a fondness for fetal tests like ultrasonography, chorionic villus biopsy, and amniocentesis, who becomes the genetic, scientific, and social father to his own monstrous, genetically engineered son, VJ.

Whereas Mary Shelley's *Frankenstein* articulates what some have called the first feminist critique of science, as well as a powerful critique of the institution of mothering, *Mutation* articulates instead a celebration of academic science.[11] The novel criticizes only the commodification of scientific procedures, not their ultimate goal: to usurp maternal procreative power. Victor Frankenstein's dabbling in alchemical arts testifies to his marginal relation to medical science, but Dr. Victor Frank's professional affiliations testify to his institutional centrality: he is acting director of the Department of Developmental Biology of Chimera, Inc., and president of its subsidiary, Fertility, Inc. The laboratory creation of his own son is merely a little "sporting with life," a sideline from those more serious and lucrative activities. During the in vitro fertilization procedure, he fuses a section of animal genetic material to the embryo's chromosomes, giving the embryo a capacity for accelerated brain growth, which he can trigger by the administration of a specific antibiotic. Implanting his embryonic son-to-be in a hired surrogate mother, Dr. Frank triggers the development of the embryo's monstrous intellect by prescribing antibiotics for the gestating woman.

VJ grows up to be just like his daddy, but worse. An embryological entrepreneur, VJ produces an entire reproductive technology laboratory, in which fifty-gallon glass tanks gestate not one, but four fetuses. In its specificity, Cook's image of babies in bottles leaves the *Sunday Mail* graphic in the dust: it almost reads like a "How to" manual of ectogenesis. The glass tanks whose sides are "fused with silicone," the heat lamps, special gestational fluid, placentas on their "plexiglass" grids with attached heart-lung machines, computer monitoring, and protein synthesizers suggest to the reader that ectogenesis is merely a matter of putting into action already existing technical knowledge. Not babies of the future, but babies of today!

And the "babies" too have an activity level resembling actual intrauterine life, as we have come to know it from prenatal ultrasound and the photographs of Lennart Nilsson.[12] Unlike the somnolent fetuses discussed earlier, Robin Cook's fetuses are quite active, "swimming about in their artificial wombs." At one stage in the narrative, they even attempt to touch their genetic mother: "Marsha timidly approached one of the tanks and peered in at a boy-child from closer range. The child looked back at her as if he wanted her; he put a tiny palm up against the glass. Marsha reached out with her own and laid her hand over the child's with just the thickness of the glass separating them" (319).

Whereas the other visual images of babies in bottles I have noted basically confine their semiotics to the realm of reproduction, this textual image ventures also into the related realms of religion and race.[13] Such allusions are not a necessary part of the pleasures *Mutation* provides for all readers, of course, but to those familiar with the cultural studies of science, Marsha's gesture of putting her hand over the child's, "with just the thickness of the glass separating them," invokes both biblical creation and interspecies contact: Michelangelo's God touches Adam's hand, and the National Geographic Society's chimpanzee touches the human hand in television specials on nonhuman primates.[14] The gesture both posits and problematizes the boundaries between "man," God, and "beast" in the era of genetic engineering and reproductive technology. It also suggests that although VJ insists on "playing God," Marsha—a psychiatrist whose role is observation and understanding rather than intervention—incarnates an alternative, female, redemptive science. With her touch, she attempts to reunite what male, instrumental science has sundered.

Commodified to both create and satisfy the desires appropriate to the genre of mass-market thriller, Cook's novel is also a rich source of cultural fantasies about reproductive technology, and it would repay further examination by readers interested in the cultural critique of science. But I leave it now via one last image of a baby in a bottle: the visual image that appears on its jacket.[15] (Ill. 4.) Product of the aptly named "One Plus One Studio," rather than of Robin Cook himself, still this image can be analyzed as part of the cultural product that is *Mutation*. The scene from *Mutation* figures ectogenesis, but this cover illustration shows a test tube, an allusion to the IVF procedure that forms part of the novel's central plot device. The "bottle" scarcely seems to contain the baby, whose folded arms extend beyond its bottom curve, but then, the baby scarcely seems to be a baby. In this jacket illustration, as in the novel, Cook challenges conventional distinctions that have anchored our understanding of reproduction both visually and terminologically, blurring them until we can't tell embryo from fetus from baby from child. (Note that the fetus in the tank is described as a "boy-child," for example.) In the text we have not one baby but four in gestational tanks; on the cover we have no baby at all, but what seems to be a boy in a bottle. With crossed arms and eyes of startling blue, a roughly four-year-old VJ stares back at the reader in a pose re-

4 Genetically Engineered Baby with His Test-Tube of Origin

markably similar to the author's photograph on the back jacket. Indeed, the composite product that is Cook's novel encourages us to connect the staring baby on the front cover with the staring author on the back. The baby in the bottle, *c'est moi*, this final image seems to say, suggesting that the very process of commodification that has shaped the new reproductive technologies has also inaugurated a new relation between author and character. The rise of the commodified literary product, it seems, has banished the romantic intersubjectivity across gender differences of Flaubert and Madame Bovary and inaugurated the technological/genetic interchangeability of Cook as author-product and his monstrous VJ.

ALTERNATIVE VISIONS FOR REPRODUCTION

What do these contemporary images of babies in bottles mean? Behind such contemporary images of reproductive technology lies a modern history that can illuminate their meanings for us and broaden our responses to the technology they represent. Scientific management of the "natural" processes of conception, gestation, and parturition has held an increasing appeal for Western culture ever since Mary Shelley's *Frankenstein*. Since the Edwardian period, when divorce rates skyrocketed and the divorce laws were rewritten, we have been increasingly fascinated—and horrified—by alternative possibilities for reproduction. Early-twentieth-century British culture featured a vigorous debate over such possibilities among birth-control reformers, sexologists, social purity campaigners, eugenicists, pronatalists, feminists, and socialists, among others. Because British culture and science exercised a hegemonic influence over American culture at the dawn of the modern industrial era, how British scientists, social thinkers, and writers spoke to these questions in the 1920s and 1930s shaped the development of reproductive technology *and* our cultural understanding of reproduction, not only in England but also in the United States. Our contemporary passion for unusual stories about reproduction —the Baby M case, children "divorcing" their parents, lawsuits over the custody of frozen embryos, reports of new high-technology methods of fertilization, debates over the benefits and dangers of genetic engineering extending to the popular culture fantasy of a Jurassic Park —all are related to the image of babies in bottles. Not only does this image call to mind the relatively recent achievement of in vitro fertilization—or test-tube babies—but it also represents a hypothetical procedure first envisioned in the early twentieth century: *ectogenesis*—or extrauterine gestation. As an image circulating throughout the history of modern representations of reproduction, then, the baby in the bottle provides the crucial, if usually unconscious, context in which we interpret, represent, respond to, and even deploy reproductive technologies today.

I may seem to be claiming an improbably large role for literature here. In part, this attention to the role of literature in the construction and representation of science reflects the new understanding of how science produces knowledge launched by Thomas S. Kuhn's *Structure of Scientific Revolutions*, with its two crucial notions, the *scientific paradigm* and the *paradigm shift*.[16] In the two decades since its publication, interdisciplinary work has increasingly linked realms and operations traditionally thought of as "literary" with other realms and operations traditionally thought of as "scientific."[17] To take an example crucial to my analysis in Chapter 1, we now understand that metaphor plays a crucial role in paradigms, functioning to "direct research and its interpretation," as Donna Haraway has argued.[18] Institutional markers of this increasing awareness of the links between literature and science are the appearance of *Configurations: A Journal of Literature, Science, and Technology* and the proliferation of conferences and learned societies, such as the Annual Meeting of the Society for Literature and Science. These new organizations reflect the increasing discursive authority of an interdisciplinary approach to science that Joseph Rouse has called the "cultural studies of scientific knowledge," including under that rubric a "heterogenous body of scholarship in history, philosophy, sociology, anthropology, feminist theory, and literary criticism." Beginning with the notion that scientific knowledge is culturally constructed, scholars in the field of cultural studies of science investigate "the traffic between the establishment of knowledge and those cultural practices and formations which philosophers of science have often regarded as 'external to knowledge.'"[19] Science is thus opened up to the wide range of analytic and investigative practices appropriate to cultural studies, ranging from context, content, and discourse analysis to analyses of the processes of production, dissemination, and consumption of scientific knowledge.

Literary forms are used to construct or consolidate a scientific field because literature is one of the institutions that shape human beings to the needs of their society through the process of identity construction that occurs in and mutually implicates both the symbolic and the material realms.[20] As Catherine Belsey has argued, representations help to construct what we understand as our cultural and social experiences, including our experiences of the body.[21] Thus literary figurations of the reproductive body have always been open to a wide range of meanings, reflecting the purposes and projects for which they are deployed. Yet as we represent reproductive experiences and procedures, we also shape them; it is my project in this book to delineate part of that (invisible) shaping process.

As Bruno Latour and Steve Woolgar explain in *Laboratory Life*, the knowledge foundational to the development of reproductive technologies, like other forms of scientific knowledge, must be transformed from disorder to order, from artifact to scientific fact.[22] And, as Allan Gross's analysis in *The Rhetoric of Science* reveals, literary forms and structures are crucial in this transformative process.[23] Analogies, metaphors, myths, and origin stories play crucial roles in the institutional construction of reproductive technology. Scientists and futurologists deploy them to promote speculation to fact, to generalize between different scientific fields and so expand the implications of scientific knowledge, to naturalize scientific operations by rewriting them into our understanding of the origins of existing relations, and to borrow the rhetorical authority of one discursive field to legitimate another.

Literary forms also come into play in scientific memoirs. This is so not only because, like all autobiographies, scientific memoirs *are* fictions—"lives" shaped by the generic expectations of life-writing—but also because scientific memoirs function as sites of explicit, retrospective construction of a scientific field. Whereas popular science writing often rewrites myth to naturalize a new scientific field, as in the To-Day and To-Morrow series, which I discuss at length in Chapter 2, scientific memoirs often produce scientific origin stories that elevate the work of scientists to the status of myth. I discuss three scientists' memoirs—Julian Huxley's *Memories*, James D. Watson's *Double Helix*, and Robert Edwards and Patrick Steptoe's *Matter of Life*—as sites for the retrospective construction of various reproductive technologies as legitimate scientific endeavors or bodies of knowledge. All three reveal the continuing influence of the literary realm on scientific practice, as well as the uses of literary structures and tropes to rewrite scientific experience to fit the conventions for scientific origin stories. As Sandra Harding describes it, these scientific origin stories "claim that the discoveries of modern science reflect the pinnacle of human progress, and that the progress science represents is lodged entirely within the scientific method. They tell us 'who we are': people who use scientific rationality to achieve progress in social life—including, of course, in inquiry."[24] Harding argues that it can be productive to read these scientific origin stories the way we read folktales and autobiographies: as symbolic expressions of a constructed reality; as selective and crafted reports limited by the self-interest, insufficient self-consciousness, and imaginative narrowness of the scientist-memoirist.[25] Rather than accepting these scientific memoirs as accurate reports of how scientific knowledge is produced, I read them "symptomatically," for what they reveal about the unconscious process of construction at work in the scientist's practice and the literary influences shaping the tale of scientific progress.[26]

THE HORIZONS OF RT

Since the birth of Louise Brown, the first test-tube baby, reproductive technology has been constructed in the popular imagination as a revolutionary new field. This construction is exemplified by the jacket copy of the medical memoir of Drs. Robert Edwards and Patrick Steptoe, who developed IVF: "On July 25th 1978, Louise Brown, the world's first test-tube baby, was born. This dramatic medical breakthrough, which was hailed in every country throughout the world, was the climax of ten years of painstaking research and often heartbreaking trial and error by two doctors—gynaecologist Patrick Steptoe and scientist Robert Edwards."[27] Hypostatizing a new era, a revolution or dramatic shift in reproductive practices caused by the intervention of modern science, Drs. Edwards and Steptoe trace the history of the test-tube baby back only as far as 1968. Such a construction of reproductive technology as a dramatic change from "traditional" "natural" methods of conception and birth obscures the complex cultural negotiations that accompanied the inception and institutionalization of this medical field, from its roots in the romantic period through its consolidation in the first four decades of the twentieth century. There is a modern history to our debate over reproductive technology, a history interweaving literature and science, a history profoundly gendered, a history of choices and struggles that we repress to our cost when we accept the contemporary construction of reproductive technology as a scientific breakthrough without a past.

A group of British scientists and writers was central to the rethinking of sexuality and reproduction—as members of such diverse organizations as the Eugenics Society; the World League for Sexual Reform; and the sexology, pronatalist, and contraceptive-education movements—in the first four decades of the twentieth century. As prominent members of the scientific and literary communities of early-twentieth-century Britain, zoologist Julian Huxley, physiologist-geneticist J.B.S. Haldane, and novelists Charlotte Haldane, Aldous Huxley, and Naomi Haldane Mitchison assessed the social, cultural, and scientific implications of scientifically mediated conception, gestation, and birth. The writings of this group reflect the mutually constitutive relationship between literary and scientific discourses that laid the foundation for reproductive technology. Long before IVF became a medical reality, and even before the development of artificial insemination (one of the earliest of the reproductive technologies to achieve popular acceptance), this group of kin and near kin wrote stories and essays that negotiated new understandings of human sexuality, gender, and procreation.[28] They wrote textbooks and essays for lay audiences on the themes of reproductive science and in novels, short stories, and a play explored the social changes such technological developments would herald.

There are two interpretive horizons for these texts and the lives of those who wrote them: a global horizon and a local one. The former is the history of representations of scientific intervention into reproduction that extends back to Mary Shelley's *Frankenstein*; the latter the more immediate context of modernist negotiations with the meaning of sexuality and reproduction in Britain from the 1920s through the 1940s. I will move from the first context to the second, with apologies in advance for a certain degree of overlap in my treatment of the modern period.

Mary Shelley's *Frankenstein* inaugurates the romantic movement of reproductive representations with its image of autonomous male birth. Written when female procreative power was being co-opted metaphorically to represent the new male birth of fraternal contractual democracy, *Frankenstein* figures as monstrous the male monopoly on political creation. Shelley's monster, as a social outcast, contests the inclusive, egalitarian pose of the liberal civil state.[29] Moreover, as feminist critics have recently shown, *Frankenstein* also operated as a powerful critique of the newly revised institution of mothering.[30] This feminist interpretive strand has its fullest elaboration in Anne K. Mellor's superb analysis of the novel, which demonstrates how its feminist critique of science and its critique of the institution of mothering converge in a nightmare image of scientific procreation that anticipates IVF.

A new scientific intervention in reproduction did appear around the time that Shelley's novel was published; the fields of embryology and obstetrics authorized new representations of the pregnant woman and the fetus harmonizing with the new political arrangements of fraternal contractual democracy.[31] Late-eighteenth- and early-nineteenth-century embryology affirmed the theory of epigenesis (the notion that an embryo develops from lesser to greater organization in the course of gestation) over the earlier theory of preformation (the notion that the embryo is a static, preformed, miniature entity somewhat like a homunculus). This shift in scientific knowledge joined Rousseauean notions of child-rearing to produce an individual fitting the needs of bourgeois capitalism. So, as Andrea Henderson has observed, "Early nineteenth century epigenesis sketches us a picture of the Romantic fetus [as] the perfect bourgeois subject—it makes itself, and so is neither simply the inheritor of paternal power nor the commodity-like product of its mother's labor"(112–113).

Parallel to the victory of epigenesist embryology was a shift in the representation of the gestating and childbearing woman in anatomical engravings and midwifery manuals, Henderson argues. Emphasizing the bony structures of the maternal pelvis as objects to be manipulated by equally rigid obstetrical instruments, these illustrations articulated a "trend . . . to present childbirth as a mechanical process, having affinities with mechanical production, but with the role of the woman . . . in the productive process . . . not as laborer but only as a machine" (103).[32] Like *Frankenstein*, these nonliterary representations participated in the romantic reconstruction of human reproductive subjects: woman, man, fetus. They reshaped the fetus as the state's ideal, organically developing, autonomous individual; they marginalized woman, exiling her from the public realm of the newly forged social contract to the private realm of the sexual contract, and they reconstructed man as both father and mother of the new political order.[33]

Modern representations of scientific intervention into reproduction built on the romantic separation of developing fetus from machine-like mother to serve ends initially not so much political, as industrial. The rationalization of labor carried on in early-twentieth-century Europe, England, and America aimed at maximum efficiency by fragmenting the work process. Treating the worker's body as a machine, the new industrial methods of Taylorism and Fordism divided the labor process into its smallest possible units, used the assembly line to enforce a uniform external schedule, and maintained constant surveillance on worker performance. Modern literature drew metaphorically on this monitored, mechanistic, regulated, and fragmented way of life, using it to figure not just production but reproduction. Drawing their central metaphor from these new techniques of industrial rationality, such literary representations imaged a scientifically reconfigured human body—both male and female—available for industrial production.[34]

LOCAL FORMS FOR RECONFIGURING THE BODY

Although the notion of reconfiguring the human body through reproductive technology had the global aim of consolidating the liberal civil state and providing docile workers for industry, in the modern period it took certain specific local forms that provide the more immediate context for this book. These projects for reimagining sexuality and reproduction, occurred at a wide variety of public sites: from organizations, to conferences, to publications. I discuss these different projects and sites at some length in the chapters to follow and will now just map them briefly in order to establish the background against which I read the texts and lives of the Huxleys and Haldanes.

The most ambitious of the organizations was the Eugenics Education Society, founded in 1907. Counting Julian Huxley, Aldous Huxley, and J.B.S. Haldane among its members, the Eugenics Society (as it was later renamed) dedicated itself to the "improvement" of the human species through a range of programs of both positive and negative eugenics—that is, through both fostering childbearing by those deemed (socially and physically) "fit" and discouraging childbearing by those deemed "unfit."

As Daniel Kevles has masterfully documented, the eugenics movement was a contested field, populated by outrageous pseudosciences as well as such legitimate endeavors as genetics and biology. Its membership was diverse as well, divided by the time of the First World War into two groups: mainline eugenics, which was identified with the poor science, racism, and classism of the Eugenics Society proper, and reform eugenics, which was the purview of Haldane and the Huxley brothers. Reform eugenicists were united in the conviction that "biology counted . . . not only did nurture figure in the shaping of man but so, significantly, did nature."[35] Unlike mainline eugenicists, however, they did not privilege biology or heredity over other explanations for the level of human achievement, such as the quality of medical treatment, food, housing, and education. Indeed, many of them viewed improvement of the human "stock" through eugenic measures as merely a necessary precondition for what they saw as more important social reforms.[36]

Sexuality and motherhood were hotly contested sites in the 1920s and 1930s, and several organizations attempted to play a part in their rethinking. Those years saw the growth of the birth-control movement, although it was "then mainly thought of as family spacing and helping in the emancipation of women, not as population control, still less as allowing general 'permissiveness,'" as Naomi Mitchison recalls in her memoir, *You May Well Ask*.[37] Dr. Marie Stopes, a long-time friend of the Haldanes, was received in the parental home by Naomi's father, John Scott Haldane, as a paleobotanist rather than the contraceptive agitator she was. "If he ever heard her name linked with less acceptable subjects he paid no attention" (34). Spurred on by Dr. Stopes's example, a number of birth-control clinics sprang up in London in the 1920s, and Naomi Mitchison served on the board of the one in North Kensington, as well as availing herself of its services to have a "dutch cap," or diaphragm, fitted. She also composed a children's play to aid in its fund-raising effort, and along with several of her friends "volunteer[ed] for tests and information . . . [and] helped at the clinic with interviews and filling in forms" (34). These organizations for contraceptive and sex education and sexual reform found common ground with the new field of sexology in 1929, when the Third International Congress of the World League for Sexual Reform met in London.

Several publications in which the Huxleys and Haldanes were active also testified to the increased interest in the scientific reconstruction of sexuality and reproduction. In April 1928, with funding from Sir Alfred Mond of Imperial Chemical Industries, a new journal was launched: *The Realist: A Journal of Scientific Humanism*. Its literary editor was Gerald Heard, whom Naomi Mitchison described in 1932 as "very nearly my best friend," and who was later to be intimate friends with Aldous Huxley during his Hollywood years.[38] The editorial board included Aldous and Julian Huxley, Naomi Mitchison and J.B.S. Haldane, as well as a clutch of other distinguished writers, scientists, and social thinkers, among them Arnold Bennett, Bronislaw Malinowski, Rebecca West, and Herbert Read. Deliberately, even defiantly interdisciplinary, the *Realist* proclaimed in its first editorial, "if Science is to become articulate (and it can hardly be humane until it does) it must learn to express itself. We stand for making the specialist understood, for introducing the laboratorist, who has lived too long with symbols, to letters."[39] Although short-lived, the *Realist* was an arena where active debate could take place between the cultures of literature and science.

Charlotte Haldane, the only member of the group who was not involved with the *Realist*, herself took the helm of a journal dedicated to bringing science to a popular audience roughly a decade later. In 1939, Haldane became the editor of *Woman Today*, a socialist journal with a modest circulation in Left Bookshops and by subscription.

Whereas her sister-in-law Naomi Mitchison was active in the contraceptive movement, publishing *Comments on Birth Control* as a Criterion Miscellany pamphlet in 1930, Charlotte's interests were engaged by pronatalism.[40] As I will discuss at greater length in Chapter 3, *Woman Today* combined the discourse(s) of science with the agendas of leftist feminism toward the goal of empowering women as mothers.

Two years after the founding of the *Realist* (at roughly the time of its demise), the proceedings of the Third International Congress of the World League for Sexual Reform were published, edited by Dr. Norman Haire. *Sexual Reform Congress* (1930) included in its prefatory matter congratulatory letters from Radclyffe Hall, Aldous and Julian Huxley, Hugh Walpole, H. G. Wells, Margaret Sanger, Jakob Wasserman, and even Freud himself. It included speeches by a wide range of writers, scientists, doctors, and reformers concerned with sexuality and reproduction, including novelists Naomi Mitchison and Vera Brittain, and sex education and contraceptive educator Dr. Marie Stopes.[41] Kegan Paul, Trench, Trubner, and Company, the forward-looking press that published Haire's book, drew on the Congress participants to expand its To-Day and To-Morrow series, a line of futurological pamphlets explicitly designed to explore issues currently under contestation, and thus to "provide the reader with a survey of numerous aspects of most modern thought."[42] Vera Brittain's talk for the Sexual Reform Congress, "The Failure of Monogamy," was reissued in the series as *Halcyon, or the Future of Monogamy*. As I will explore in greater detail in Chapter 2, it mapped out a promising feminist response to the notion of ectogenesis featured in the premiere volume of the series, J.B.S. Haldane's *Daedalus, or Science and the Future* (1923).

As all of these organizations, congresses, and publications reveal, reproductive technology as both image and concept is *malleable*, deployable differently and to different ends, depending on its context. We can read the image of reproductive technology to illuminate the ideology or ideologies in the process of cultural construction in an era, as Mary Poovey asserts in the epigraph to this introduction. But the history of representations of reproductive technology, the story of how they shaped and were shaped by its development in the early years of this century, can also illuminate our contemporary moment. I work between the modern and the postmodern moments in the understanding that each can illuminate the other. My aim is to show how the modern imaginative construction of reproductive technology helped to shape our postmodern practices, as well as how a familiarity with modern images, fantasies, practices, and narratives of the scientific intervention in reproduction can bring back into our awareness issues that are repressed or denied by our contemporary construction of reproductive technology.

DISJUNCTION IN THE FEMINIST RESPONSE TO RT

I have long been interested in the contemporary cultural prominence of images of reproductive technology, particularly in the works of women writers and feminist theorists. I have wondered what work of ideological construction was being carried out through the production and dissemination of those images. I have been fascinated by the rich and diverse literary representations of this new medical-scientific field, most notably in the works of Joanna Russ, Marge Piercy, Margaret Atwood, Octavia Butler, Elizabeth Jolley, Angela Carter, and Fay Weldon.[43] But I also noted a puzzling disjunction between the emancipatory interpretations of reproductive technology in many, though not all, of those novels and in some of the literary critical writings on them, and the negative responses of feminist theorists and activists to the actual implementation of those technologies in Europe and North America.[44] Although many postmodern literary critics praised the potential of these technologies to destabilize gender identities and call into question the boundaries of self, society, even species, I found that in the same historical moment many feminist theorists and activists were indicting the actual uses of those same technologies as "unsuccessful, unsafe, unkind, unnecessary, unwanted, unsisterly, and unwise."[45]

To some degree, this gap between praise and blame reflects the chameleon nature of the technologies themselves. As N. Katherine Hayles has pointed out, reproductive technology is a preeminent example of the denaturing of the human body that characterizes cultural postmodernism: "When the genetic text of the unborn child can be embedded in a biological site far removed from its origin, the intimate connection between child and womb which once provided a natural context for gestation has been denatured."[46] The Janus-faced quality of cultural postmodernism, as Hayles anatomizes it, can help to explain the representational gap between contemporary literature and literary criticism on one side and contemporary feminist theory and praxis on the other. Hayles distinguishes between theoretical and technological postmodernism: "In its theoretical guises, cultural postmodernism

champions the disruption of globalized forms and rationalized structures. In its technological guises, it continues to erect networks of increasing scope and power"(291).

This distinction plays itself out in the different sites where reproductive technology operates, whether in representation or in "reality." In the realm of theory, reproductive technology can function as
emancipatory, enabling Octavia Butler to imagine a more complex origin story, and thus a more complex subject position, for her multi-
parented alien, Akin, in *Imago*. But reproductive technology can also
be used for acts of oppression ranging from racist eugenics measures
to female feticide. And these alternative possibilities are not free, but
mutually determined, indeed mutually reinforcing: as Hayles observes
in *Chaos Bound*: "Despite their apparent opposition, these two aspects
of cultural postmodernism engage each other in self-sustaining feedback loops" (291).

In addition to the broad subject of the divided response of contemporary feminists to reproductive technology, I am interested in a more
specific contemporary site where reproductive technology is contested, because such a site would exemplify the central role played by
aspects of literary discourse in the construction of scientific facts. This
is the British debate (1982–1990) over reproductive technology and,
more specifically, research on human embryos. Because I draw on this
debate frequently for the questions or issues that frame my examination of the Huxleys' and Haldanes' reproductive writings in the 1920s,
1930s, and 1940s, I want to sketch its general parameters here.

This debate began in 1982, when the Committee of Inquiry into
Human Fertilisation and Embryology, commonly called the Warnock
Committee, was established in response to the birth of Louise Brown
in 1978. This group of physicians, scientists, and social scientists, ethicists, and lay people was charged by the British government "to examine the social, ethical and legal implications of recent, and potential
developments in the field of human assisted reproduction."[47] The
Warnock Report, as it came to be called, was issued in the summer of
1984. It contained a lengthy, painstaking review of the issues raised by
reproductive technologies including infertility treatments (AID, IVF,
egg and embryo donation, and surrogacy); the wider uses of these
techniques to prevent the transmission of hereditary diseases, for the
purpose of gender identification including post-fertilization and pre-
implantation gender identification, and for sex selection; the freezing
and storage of human semen, eggs, and embryos; the scientific issues
raised by such techniques (research on human embryos); possible fu-

ture developments in research (fertilization across species, the use of human embryos for drug testing, ectogenesis, the gestation of human embryos in other species, parthenogenesis, cloning, embryonic biopsy, nucleus substitution, and therapies to prevent genetic defects); and strategies for regulating both research and infertility services.

The Warnock Committee recommended the establishment of a new "statutory licensing authority" to regulate both infertility services (AID, IVF, egg and embryo donation, and the "sale or purchase of human gametes") and research; they called for the creation of legal limits on research; they advocated the continued and widened provision of infertility services through the National Health Service; and they recommended a series of legal changes to regulate infertility services and to regularize the legal and social position of the children resulting from them (Warnock Report, 80–81). However, there were two areas in which the Committee could not agree: surrogacy and embryo research. Minority reports in those areas left the issues open to public debate. Six years later, the Human Fertilisation and Embryology Bill was introduced in Parliament, and the whole range of issues in that debate had extensive airing, both in Parliament and in the popular media. The act, passed in 1990, permits embryo research to continue up to fourteen days after fertilization and establishes "a statutory body which would issue licenses for the use of embryos in the treatment of infertility, congenital disease, miscarriage, the development of contraceptives including a vaccine, and more effective methods of detecting chromosomal abnormalities."[48]

The "Embryo Bill" debate produced what Marilyn Strathern calls a "cultural education" as people encountered "these issues for the first time and largely via the popular media (press, television, public conferences)."[49] Strathern examines what the tenor of the debate reveals about the unconscious and habitual notions of kinship, family, nature, and culture held in Europe and North America, in part by deliberately making the juxtaposition to her own research field: Melanesian ethnography. My interests are somewhat different from Strathern's: her approach is cross-cultural, mine might be better described as transhistorical. I read the parliamentary debate, as it was recorded in the Official Reports of the House of Commons and House of Lords, and as it occurred and was documented in the popular media, with an eye to the metaphors, images, and representations of reproduction deployed there. Taking these metaphors, images, and representations not as natural and inevitable, but as ideologically constructed, I ask what relation they have to the earlier debates over reproductive technology occurring in the British popular media and popular science publications from the 1920s through the 1940s. Despite my different emphasis and focus, however, I am indebted both to Strathern's exemplary deployment of what Latour and Woolgar have called the technique of anthropological strangeness—choosing to apprehend as strange those aspects of a culture or activity customarily taken for granted—and also to her concept of domaining.[50]

Fundamental to my approach in this book is the understanding that our images of reproduction are shaped by the arena of ideological construction or contestation within which they come into being. Moreover, images, and the ideas they embody, change when they move between discursive communities. This is what Marilyn Strathern has called the "domaining effect": that "in cultural life . . . the ideas that reproduce themselves in our communications *never reproduce themselves exactly. . . .* [I]deas are always enunciated in an environment of other ideas, in contexts already occupied by other thoughts and images" (*Reproducing the Future*, 6). My opening discussion of the images of babies in bottles reflects the basic tenet of domaining: that the same reproductive image can have very different implications depending on the domain within which it appears and the ideological work that it is doing.

We can see the impact of the domaining effect if we return one last time to the images of babies in bottles with which I began: three of them visual (from *New Age*, the *Sunday Mail*, and Dobyns's review of *Geek Love*) and one textual (*Mutation*). *New Age* is a magazine for the middle-class consumer; appropriately enough, it represents the baby in the bottle in terms of the implications of reproductive technology for the autonomous individual first, the collectivity only second. In contrast, as befits a working-class tabloid, the *Sunday Mail* image reveals a concern for the possibility that science can be used to promote race- or class-based oppression, as well as to improve the lot of the ordinary worker. *Geek Love* is a novel. The illustration accompanying its review is shaped (appropriately, as it happens) by the expectation of a plotline: we become curious about the hand pulling back the curtain, the babies pickled in the jar, the person who wrote the label. Finally, *Mutation* is a mass-market thriller. Its textual image of babies in bottles reflects the complex and self-contradictory imperatives of the genre, combining believable technical details with lurid hyperbolic imagery. So, Cook gives us not one "baby" but four, and a "bottle" emitting the "eerie blue light" of the horror film, but with instrumentation possessing the scientific gloss of the most up-to-date industrial laboratory.

As this final reading of the four images of babies in bottles reveals, different meanings are conveyed by these different representations of reproductive technology depending on their different discursive and ideological contexts. Yet as icons mediating between science and the lay community and possessing both aesthetic and scientific meaning and value, these images also harken back to images from an earlier time: the ectogenetic fetuses and other such reproductive constructions of the 1920s and 1930s. Those early-twentieth-century images of reproductive technology, like their late-twentieth-century progeny, bear the traces of complex negotiations over issues of genre and gender, race and class, and literary and scientific values.

In the chapters to come, I consider the literary and popular science writings of Julian Huxley, J.B.S. Haldane, Charlotte Haldane, Aldous Huxley, and Naomi Mitchison—writings that contain a range of representations of reproductive technology from babies in bottles to surrogate mothers. In each chapter, I focus on the writing of one of these scientists and/or novelists to illustrate some of the representational history of reproductive technology, as well as to illuminate current questions in reproductive technology.

In these texts representing reproductive technology, the Huxleys and Haldanes articulated the ideological and representational forces that helped to shape reproductive technology. In particular, they dramatized the gendered construction of the modern scientific project and the contrasting and complementary representations of reproduction articulated by literature and science in the early twentieth century. By considering how early-twentieth-century fiction and popular science writings negotiated the issues central to the project of achieving human control over reproduction, we can reclaim the origins of this postmodern technology. It is to those earlier images that we must look if we want to understand what acts of ideological construction have been carried out, and are currently being performed, in the name of reproductive technology.

one

Babies in Bottles and Tissue-Culture Kings

THE ROLE OF ANALOGY
IN THE DEVELOPMENT
OF REPRODUCTIVE TECHNOLOGY

Analogy is in the majority of cases the clue which guides the scientific explorer towards radically new discoveries, the light which serves as the first indication of a distant region habitable by thought. —JULIAN HUXLEY, "Science, Natural and Social"

Modernist knowledge practices themselves . . . afford analogies for the way differentiations are endlessly repeated . . . and are recreated every time one tries to "know" something. For the difference between orders of knowing (recognition, construction) is a relation integral to such knowledge. . . . It is the misfortune of interpretation, like culture, to appear as artificial as analogies and metaphors always do in this system.

—MARILYN STRATHERN, *Reproducing the Future*

In 1990, Parliament passed the Human Fertilisation and Embryology Act, which authorized research on human embryos up to fourteen days old. One of the major issues raised during the debate was whether or not embryological development occurred in discrete stages, and thus whether a specific stage could be identified at which an individual identity was formed. In the course of the debate, the archbishop of York put forward a remarkable analogy to ground his argument that no such threshold of human identity was known or knowable: embryonic development, he argued, is like a Mandelbrot set, the intricate, infinitely regressing shape that maps the behavior of dynamical systems.[1] Discovered in 1980 by Benoit Mandlebrot of IBM, the Man-

24

dlebrot set is a computer-generated image that represents the mathematical operation yielding "all choices of c that can stay bounded."[2] Certainly, the invocation of chaos theory in a debate on human embryo research is surprising, but what interests me most about this episode is how it problematizes the operation of analogy itself. How does an analogy inform our representation of reproductive technology?

In her perceptive study of the debate over the Human Fertilisation and Embryology Bill, Marilyn Strathern argues that the analogy between the Mandelbrot set and embryological development epitomizes the operation of modernist knowledge practices: analogy purports to describe objects of knowledge by juxtaposing two unrelated fields, yet the very act of asserting a parallel between two discrete realms—here embryology and chaos theory—is not description but construction posing as description.

> The metaphor occupied the obvious place of a constructed act of interpretation . . . assisting interpretation by so partitioning off this "natural" domain through the artificiality of the metaphor had the effect of glossing over existing partitions between (natural) fact and (social) interpretation. The flow from "cell" to "person," like the flow from "scientific information" to "legislative decision," could be rendered continuous. (*Reproducing the Future*, 144)

Despite her awareness of the constructive function of analogies, however, Strathern dodges the full implications of the Mandlebrot set analogy when she characterizes "the metaphor so brilliantly summoned" as "an artificial graft, one that bore no intrinsic relationship to the subject of embryo development" (144).

Analogies are neither merely artificial grafts, nor wholly innocent. I suggest that we would do well to reconsider the relationship between the analogies we make and the fields they are intended to illuminate. In this case, that field is embryology. Using chaos theory to illuminate human embryology is no more "an artificial graft" than is Strathern's language in this passage. The analogy Strathern chooses is determined both by her topic (reproduction) and by her argument: that analogy is used to obscure the difference between discursive realms. Her (probably unconscious) choice of metaphors enables her to confirm her own argumentative position by not merely asserting, but also enacting, the constructive function of analogy. By describing the archbishop's analogy as "an artificial graft," she represents the act of grafting itself as neither constructing nor constructed, but natural.

A similar rhetorical strategy motivates the archbishop of York in his choice of analogy: he uses one that increases the force of his argumentative position. His Mandelbrot set analogy asserts a relationship to the broader discursive field under debate—the general realm of scientific knowledge—in order to displace or replace the specific scientific field now under discussion by another, even more authoritative scientific field. So, the field of embryology is displaced by the even more authoritative field of mathematics.[3]

I find a convergence between Strathern's argument that analogies function to construct as well as describe and Allan Gross's argument that analogy has both a heuristic, or hypothesis-creating, and probative, or proof-making, purpose in the sciences. Analogy functions, in Gross's view, as an agreed-upon realm of experience that carries out the move from hypothesis to proof, or from context of discovery to context of verification, "*as if* moving from appearance to reality."[4] As Donna Haraway demonstrates in her brilliant study of metaphors in early-twentieth-century developmental biology, analogies, like metaphors, serve to anchor paradigms, and changes in analogies and metaphors often mark shifts in paradigms. Moreover, analogy is not a neutral scientific tool: "Analogy and primary referent are both altered in meaning as a result of juxtaposition."[5] In short, analogy shapes not only how we interpret scientific findings but the scientific findings themselves: it determines the direction of scientific practice, the questions asked, the results obtained, and the interpretations deduced.

For critics interested in the relationship between scientific and literary representation, as I am, analogy is a valuable analytic tool. Analogies reflect ideological pressures, as Strathern acknowledges when she observes that "culture consists in the way analogies are drawn between things, in the way certain thoughts are used to think others" (*Reproducing the Future*, 33). Moreover, because analogy functions not only within but also between different discursive realms, we can use it to study the important phenomenon that Strathern has called "the domaining effect," the subtle shift that takes place in ideas when they move from one cultural or social context to another, since ideas "never reproduce themselves exactly" in the new realm (6). As Strathern explains this effect,

> In cultural life, in those habits of thought about which for most of the time we are very much unaware, the ideas that reproduce themselves in our communications *never reproduce themselves exactly*. They are always found in environments or contexts that have their own properties

or characteristics. . . . Moreover, insofar as each is a domain, each imposes its own logic of "natural" association. Natural association *means* that ideas are always enunciated in an environment of other ideas, in contexts already occupied by other thoughts and images. Finding a place for new thoughts becomes an act of displacement. (6)

As part of her examination of the different ways that kinship is conceptualized, Strathern compares the different uses of analogies for reproduction and identity in Melanesian and Western cultures; the domaining effect becomes apparent in the traffic back and forth between them.

Strathern focuses on the contrast between contemporary cultures, whereas my interest is, as I have said, more transhistorical. I believe that reproductive ideas circulate through the overlapping realms of literature, popular culture, and science via the operations of analogy, and that an understanding of the domaining effect, as it functioned in that circulation of ideas in Britain in the 1920s and 1930s, can illuminate our present understanding of reproductive technology. If we understand those shifts in meaning produced by the domaining effect, we will have a broader repertoire of responses to the challenge posed for us all by reproductive technologies. We can ask, How do analogies between human and animal reproduction, between individual embryological development and species evolution, shift when they travel between the realm of fiction and the realm of "the construction of scientific facts"?[6] How do they function to shape and redirect those practices, whether scientific or literary and cultural?

My test case, or point of entry, to these issues is the role of analogy in the life and writings of zoologist and popular science writer Julian Huxley (1887–1975). As he observed in the epigraph that opened this chapter, Huxley felt analogy was essential to the scientist's work, shaping not only the answers found but, more important, the questions asked, and guiding the experimenter toward new areas of scientific exploration. Unreflective use of analogy has its dangers, Huxley also warned: "Analogy, unless applied with the greatest caution, is a dangerous tool" because it "may very readily mislead."[7] There is a tension between Huxley's awareness of how such cultural practices as analogizing can shape scientific practices and his inattention to the ideological implications of his own use of analogy. Drawing analogies between human and animal reproduction, between individual and species development, Huxley's work also exemplified the domaining effect that occurs when reproductive ideas are applied to different discursive fields. As we will see, when reproductive ideas are shifted from the realm of fiction to the milieu of experimental science, they take on the instrumental perspective prevalent in contemporary scientific culture, and a concern with the implications of outcome is replaced by a focus on the details of method.

In a variety of personal, popular-scientific, and fictional writings, Huxley's analogies not only reflected but actually determined the construction of the scientific intervention into reproduction that would come to be known as reproductive technology. Although he used analogy to broach the topic of reproduction in a remarkably early letter to his grandfather, he gave the topic extensive consideration in his popular science writing of the 1920s, his memoirs, and his essays on eugenics. Still, his science fiction short story, "The Tissue-Culture King" (1926), stands as his most illuminating text about reproductive technology. His first venture into fiction, it took as its explicit subject the social effects of four reproductive technologies: "tissue culture; experimental embryology; endocrine treatment; [and] artificial parthenogenesis."[8]

These reproductive technologies are familiar to us today either because they are in active medical use or because they have received extensive treatment in fiction. Endocrine treatment, experimental embryology, and tissue culture are essential techniques in contemporary in vitro fertilization. To summarize their role in that procedure: The woman begins with a course of fertility drugs (Clomiphene, follicle-stimulating hormone, human chorionic gonadotrophin) to stimulate the production and ripening of eggs; gametes are then collected and mixed (at a ratio of one egg to about 50,000 sperm cells) outside the human body in the familiar test-tube or petri dish; this mixture is incubated under the supervision of embryologists in a tissue-culture medium, in which the fertilized embryo undergoes cell divisions until it is an eight- to sixteen-cell embryo (approximately forty-eight to seventy-two hours); then the embryo is transferred to the uterus of the gestating woman and gestation proceeds normally.[9] Artificial parthenogenesis, the fourth reproductive technology featured in Huxley's story, is still only a hypothetical reproductive technique in human beings; however, it has, since the turn of the century, been a popular theme for feminist fictions from Charlotte Perkins Gilman's *Herland* (1915) to Fay Weldon's *Cloning of Joanna May* (1989), a blithe thriller of "parthenogenesis plus implantation, and a good time had by all."[10]

Those technologies had a different meaning in Huxley's day, when they were either under initial exploration or still hypothetical, than they do in our own, when all but parthenogenesis is an accomplished medical-scientific procedure. In order to appreciate Huxley's representation of these technologies in "The Tissue-Culture King," we need to establish not only what those technologies mean now, but also how they were understood by scientists in the 1920s and 1930s. Moreover, since those reproductive technologies took on a different meaning in fiction than in scientific discourse because of the effect of "domaining," we need to consider the fictional as well as the scientific context for their representation.

BABIES IN BOTTLES

In 1926 the thirty-eight-year-old Julian Huxley wrote, "Development is to the individual what evolution is to the race."[11] In March 1892, the four-year-old Julian Huxley read Charles Kingsley's children's story *The Water-Babies* (1863) and his interest was captured by the illustration "representing [his] grandfather [T. H. Huxley] and Professor Owen examining a bottled water-baby with magnifying glasses." (Ill. 5.) He wrote a letter to his grandfather: "Dear Grandpater have you seen a Water-baby? Did you put it in a bottle? Did it wonder if it could get out? Can I see it some day? Your loving JULIAN."[12] His grandfather replied by return post: "My dear Julian I never could make sure about that Water Baby. I have seen Babies in water and Babies in bottles: but the baby in the water was not in a bottle and the Baby in the bottle was not in water."[13]

A genealogy of Julian Huxley's embryological and reproductive writings might well begin with this encounter with Charles Kingsley's *Water-Babies*, for Huxley repeatedly referred to that Victorian story in the adult writings in which he articulated his own, distinctively modern, understanding of individual development and species evolution. Huxley's letter to his grandfather deals with themes to which he would return with almost obsessive frequency as a professional zoologist and popular science writer: embryonic development ("water-babies"); desire to make something visible ("Did you see a water-baby?"); scientific power and control ("Did you put it in a bottle?"); conflict between empathy and scientific objectivity ("Did it wonder if it could get out? Can I see it some day?"). Despite its playfulness, his grandfather's response explored serious issues too, testing the boundaries of life and death, of woman's body and machine, of fiction and fact—themes that would

Well.   How do you know that somebody has not?

" But they would have put it into spirits, or into the *Illustrated News*, or perhaps cut it into two halves, poor dear little thing, and sent one to Professor Owen, and one to Professor Huxley, to see what they would each say about it."

Ah, my dear little man! that does not follow at all, as you will see before the end of the story.

" But a water-baby is contrary to nature."

Well, but, my dear little man, you must learn to talk about such things, when you grow older, in a very differ-ent way from that.   You must not talk about " ain't " and " can't " when you speak of this great wonderful world round you, of which the wisest man knows only

5 The Original Baby in the Bottle, from Kingsley's *The Water-Babies*

also reappear in Huxley's later work. Although at different times in Huxley's life these issues were mediated by different discourses—ranging from fairy tale to autobiography, science fiction, and popular science writing—they were most often approached analogically via the theme of reproduction, the site for the individual and collective production and management of difference.

Two characters from *The Water-Babies* play a particularly striking role in Huxley's adult scientific understanding of reproduction: (1) *the figure of the water-baby*, which Kingsley parallels to the eft in a densely interconnected representation of human biological and moral development with species evolution, and (2) *the figure of Mother Carey*, who "makes things make themselves."[14] These figures anticipate the two major strands of Huxley's scientific career: his commitment to understanding individual human development in the broader context of the human species—and the whole biological, multispecies world (identity-as-analogy)—and his interest in exploring the processes by which sexual reproduction and embryological development produce an individual (identity-as-differentiation). They also anticipate "The Tissue-Culture King" in ways that will become apparent later.

The parallel between the water-baby and the eft reflects Kingsley's appreciation for "the enlargement of kinship—the great family which must . . . include the chimney-boy and the scientist, and which moralises the connections between plants, animals, and human life."[15] The transformation of Tom the chimney sweep into a water-baby yokes a moral tale of Wordsworthian life before birth to the scientific narratives of embryology and zoology.[16] Tom tumbles into the stream and finds himself "about four inches, or—that I may be accurate—3.87902 inches long, and having round the parotid region of his fauces a set of external gills . . . just like those of a sucking eft" (Kingsley, *Water-Babies*, 49). "Quite amphibious," Tom straddles the species' boundary, and his story enables Kingsley to play with the interwoven narratives of human biological, moral and species development (62).

Kingsley's story straddles a discursive boundary too, as a multilayered drama of evolutionary, embryological, and moral development. In a passage proving the existence of water-babies to a hypothetical skeptical reader, the narrative explicitly urges the reader to connect these disparate discourses by the operation of analogy:

> If he says (as he most certainly will) that these transformations only take place in the lower animals, and not in the higher, say that that seems to little boys, and to some grown people, a very strange

fancy. . . . And if he says (as he will) that not having seen such a change
in his experience, he is not bound to believe it, ask him respectfully
where his microscope has been? Does not each of us, in coming into
this world, go through a transformation just as wonderful as that of a
sea-egg, or a butterfly? and does not reason and analogy, as well as
Scripture, tell us that that transformation is not the last? and that,
though what we shall be, we know not, yet we are here but as the crawl-
ing caterpillar and shall be hereafter as the perfect fly. (55)

The shock of Darwinian theory prompted Victorian writers to "nat-
uralise the new theories back into creationist language," Gillian Beer
has observed (*Darwin's Plots*, 129). In its unself-conscious intermingling of religious, evolutionary, zoological, and embryological sites
of transformation, Kingsley's proof of water-babies does indeed seem
to reveal naturalization in process: fiction (the existence of water-
babies) is confirmed by fact (the existence of efts). Yet what we have
here is finally less a process of naturalization than a chain of analogies:
Darwinian evolutionary theory is like human development; human
experience is like scientific data; human embryology resembles developmental zoology; and developmental zoology recalls scriptural truth.
That the relative truth claims of these different categories of knowledge
were still in flux at the time of *The Water-Babies* (in contrast to Julian
Huxley's time, or our own) is apparent if we ask ourselves which term
in the chain of analogies authorizes the whole, "scriptural" or "empirical scientific"? With religious and scientific knowledge still contesting
dominance in Kingsley's day, we may conclude that it is the hybrid figure of the water-baby himself, midway on his moral-biological transformative pilgrimage, that authorizes the complex linkage between
two disparate realms: the spiritual and the scientific.

The parallelism between moral growth and scientific education is
unconsciously racist—an outgrowth of a proto-eugenic belief in evolutionary hierarchy. In his travels in the waters, Tom is introduced to all
manner of reproductively tinged metamorphoses, enforcing the Victorian values of kindness, modesty, industry, responsibility, and fidelity
while underscoring the almost limitless diversity of such biological categories as species, age, and sex. Yet if sex, age, and species are fluid for
the water-babies and the other nonhuman denizens of the waters, for
human beings sex, race, and class generate more fixed, and mutually
reinforcing, social positions. Sooty and ill-mannered at the outset
(taken by the "little white lady" to be a "little black ape"), Tom learns to
ascend the moral ladder and, as he does so, finds himself rising on the
evolutionary and racial ladders as well (19). When Tom has an out-

break of "naughtiness," for example, he undergoes devolution, to find himself "all over prickles, just like a sea-egg" (161). (Ill. 6.)

Biological metamorphosis is paralleled by moral, social, and epistemological transformation under the educating influence of the utilitarian pair Mrs. Bedonebyasyoudid and Mrs. Doasyouwouldbedoneby, until Tom becomes "a great man of science, [who] can plan railroads, and steam-engines, and electric telegraphs, and rifled guns, and so forth . . . " (243–244). Tom's evolutionary ascension is complete when he is permitted to "go home with Ellie [the 'little white lady'] on Sundays, and sometimes on week-days, too" (243). Although they do not marry, for "no one ever marries in a fairy-tale," the implication is that the sooty little poor boy has now become—by dint of hard work and good works—a fitting mate for the little rich white girl (244). And, married or not, since they have mated, moral education will soon produce species evolution, the conclusion implies.

The figure of Mother Carey, like the image of the water-baby, anticipates the concerns of Julian Huxley's adult scientific work: individual development and species evolution. But whereas the water-baby invokes the embryo or fetus, icon of individual biological development, Mother Carey invokes the collective social and biological development of the species, fueled by the twin doctrines of Darwinian evolutionism and imperialism. For Mother Carey resembles Queen Victoria, ruling over a diverse and extensive empire that stretches across the seas: "[She was] the grandest old lady [Tom] had ever seen—a white marble lady, sitting on a white marble throne. And from the foot of the throne there swum away, out and out into the sea, millions of new-born creatures, of more shapes and colours than man ever dreamed" (200). The aspect of Mother Carey that young Julian Huxley seems to have remembered most is not her fecundity but her magical, invisible power to catalyze development in others.[17] This elusive power would fascinate the adult Huxley as a zoologist; in a popular science essay written in the early 1920s, he describes the process guiding the development of tendon fibers as analogous to "old Mother Carey, sitting idle all day, but making things make themselves."[18] Even from the vantage point of old age, he recalled "the power and the independence of nature— nature that helps things make themselves, as Charles Kingsley wrote in *The Water-Babies*."[19]

Kingsley's fairy tale concludes by endorsing limits to scientific curiosity. Even though Tom "knows everything about everything" once he gets to be a "great man of science," the narrator explains, Mother Carey's knowledge must still remain closed to him, questions which could not bear the sweets : but took them again in spite of himself.

And when Mrs. Doasyouwouldbedoneby came, he wanted to be cuddled like the rest; but she said very seriously :

" I should like to cuddle you ; but I cannot, you are so horny and prickly."

And Tom looked at himself : and he was all over prickles, just like a sea-egg.

Which was quite natural; for you must know and believe that people's souls make their bodies just as a snail makes its shell (I am not joking, my little man; I am in serious, solemn earnest).

And therefore, when Tom's soul grew all prickly with naughty tempers, his body could not help growing prickly too, so that nobody would cuddle him, or play with him, or even like to look at him.

What could Tom do now but go away and hide in a

6 Moral Degeneration as Species Devolution: Naughty Tom as a "Sea-Egg"

Kingsley represents as embryological and evolutionary. But as if deliberately turning his back on Kingsley's moral, Julian Huxley's contributions as a "great man of science" focus precisely on issues of development, differentiation, and evolution. In his scientific experiments, as in his popular science writings, he would disregard the warning in *The Water-Babies* that some questions must remain unanswered —especially those concerning the origin and development of life. Instead, Huxley's adult scientific work concerned the very questions Kingsley declared off-limits: embryological, evolutionary questions, such as "why a hen's egg don't turn into a crocodile" (244).

THE AXOLOTL EXPERIMENT

In 1919, having read of the work of J. F. Gudernatsch, who "introduced premature metamorphosis of tadpoles into froglets by feeding them on thyroid gland," Huxley wondered, "what would happen if I gave the same diet to axolotls?"[20] The axolotl, as Huxley described it in *Memories*, bears a striking resemblance to Kingsley's efts and water-babies: "The axolotl is a strange tailed amphibian from Mexico . . . [that] normally lives permanently as a tadpole or eft, with moist skin, external gills to breathe with, and a broad swimming fin round its long tail" (119). Feeding the axolotls minced thyroid bought from a local butcher, Huxley inflicted upon them an artificial maturation. "The gills shrank, the membrane round the tail became resorbed, and the aquatic efts turned into large salamander-like creatures with dry skin, adapted to air" (120). Watching this induced metamorphosis, Huxley exulted in its departure from nature's way of making things make themselves. As he recalled in his memoirs, "It really was exciting to have recreated a land animal which had not existed, except in tadpole form, for many thousands of years" (120).

Not only did Huxley's scientific investigations concern a creature remarkably reminiscent of Kingsley's water-babies, but his own writing about the axolotl experiment (and his wife's as well) continues the analogy, adopting Kingsley's authorial strategy of layered resemblances in *The Water-Babies*. A parallel conceptualization of efts and fetuses, zoology and human embryology, animal and human reproductive science appears in the autobiographical narratives of Julian and his wife, Juliette, at the time he carried out the axolotl experiment. In *Memories*, Huxley recounts: "In the autumn, [when] I started feeding some axolotls on thyroid . . . Juliette complained that I was much more interested in them than in her welfare and that of our first child, born on December 2, 1920" (119). And Juliette herself recalls, "During the next year [1922–1923], while Julian was finishing his *Essays of a Biologist*, Francis was quietly accomplishing his own little miracle of embryonic evolution, and was born in late August 1923, at Holywell, Oxford."[21] The passages draw implicit analogies between animal and human, ax- olotl and fetus, Julian's scientific "feeding" ("endocrine treatment") and Juliette's maternal, placental feeding, individual embryological development and species evolution, as well as literary production and human reproduction.

Journalism, too, revealed traffic between species and genres in its coverage of the axolotl experiment. When Huxley published his find- ings in *Nature*, the *Daily Mail* picked up the story, trumpeting "Young Huxley has discovered the Elixir of Life!" and, in a later story, blaring "A Great Discovery. Thyroid Gland Marvels. Control of Sex and Growth. Renewal of Youth."[22] (Ill. 7.) Blurring the boundaries of fact and fiction, myth and reality, question and answer, animal and human, the *Daily Mail* identified the "problems illustrated by a series of discov- eries . . . in laboratories at Oxford," while failing to point out that to identify a problem was not (yet) to discover a solution, whether for the tadpole, flatworm, frog, axolotl, or human being. Its opening para- graphs combined hackneyed science fiction formulas for human bet- terment with a risible list of "actual achievements" confined to amphibians and worms:

> The secret of perpetual youth and renewed vigour, the determination of sex, and the curing of certain human diseases are some of the prob- lems illustrated by a series of discoveries now being unearthed princi- pally in laboratories at Oxford. Actual achievements in the last few years include:—
> Changes of tadpoles into frogs within three weeks.
> Production of a new sort of creature.
> Restoration of a flatworm to youth.
> Control of the sex of frogs' eggs, producing 90 per cent. males at will.[23]

Illustrated by a "before and after" photograph of Huxley's axolotl, the article amalgamates the mythic, the commercial and the scientific: dreams of "perpetual youth" jostle copy recalling advertisements for self-improvement nostrums and restrained lists of far less grand ex- perimental findings. The mingled mythic and commercial prose at- tests to the readiness of the human imagination to welcome any experiment that could promise the extension of life or an increase

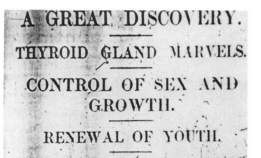

A GREAT DISCOVERY.

THYROID GLAND MARVELS.

CONTROL OF SEX AND GROWTH.

RENEWAL OF YOUTH.

The secret of perpetual youth and renewed vigour, the determination of sex, and the curing of certain human diseases are some of the problems illustrated by a series of discoveries now being unearthed principally in laboratories at Oxford. Actual achievements in the last few years include:—

Change of tadpoles into frogs within three weeks.

Production of a new sort of creature.

Restoration of a flatworm to youth.

Control of the sex of frogs' eggs, producing 90 per cent. males at will.

Two English men of science, both with famous names, have now associated themselves to hunt down the clue. One is Mr. Julian Huxley, a Fellow of New College, Oxford, one of the most brilliant of our young biologists.

MR. JULIAN HUXLEY.
["Daily Mail."]

Mr. Huxley discovered in December last that by giving a preparation of the thyroid gland (from any animal) to tadpoles he could change them into frogs in three weeks or so instead of three months odd, which is the natural period. On the other hand, by removing this minute gland they could be kept as tadpoles till he gave them thyroid, when they began to develop into frogs.

He has now done a second and more difficult transformation trick. He has given thyroid to the axolotl—a Mexican tadpole-like creature that has only two or three times in its known history developed into a sort of frog or salamander. But Mr. Huxley by his thyroid diet has easily per-

Left: The Axolotl in its ordinary state. Right: After feeding with the thyroid gland. The alteration in the gills and head is apparent.

a biologist has taken the marvel a step further. He has concocted in his laboratory a chemical that is similar to the thyroid; and this chemical—chiefly iodine—can bring about very much the same transformations as does the living gland. A mere chemical can bring about a vital change.

An American, a Danish, and a Czech biologist have independently and nearly simultaneously arrived at similar results with the thyroid gland—though not with its chemical imitation. An American has actually restored a certain rudimentary animal known as the flatworm to youth; and it has lived in this recovered youthful state through 18 generations of worms descended from its sister.

DOUBLED ENERGY.

The astounding effect of applications of thyroid gland in certain human diseases, especially on cretinous (stunted in mind and body) children, has been known for some time. The whole face can be changed by its application, as well as the mind and health. But with this application Mr. Huxley is not directly concerned, though doctors will follow his researches with close attention.

It has been found that the speed of vital processes of a man may be increased several per cent. by an application of no more than the 30,000th part of an ounce of this chemical diet into mind and youth.

Mr. Huxley is a son of Dr. L.

7 The Axolotl Experiment Makes Headlines

in sexual prowess. That there was a huge market here—for information and advice, if nothing more—was made clear by the scale of response to the extensive publicity given Huxley's experiment: "I was bombarded with letters from cranks and sufferers from all over the world. One pathetic writer from India lamented that he possessed 'an undersized and under-developed male organ' and demanded to know whether it could be 'at least doubled in dimensions' by this miraculous thyroid treatment" (120). Teased by J.B.S. Haldane that such publicity would destroy his standing as "a reputable scientist," Huxley hurried to write an article "clearing up the facts," and thus launched his career as a popular science writer."[24]

The episode of the axolotl experiment suggests several things about how different knowledge practices shape our understanding of reproduction. First, it reveals the power of analogies to guide our thinking in a variety of different fields. Huxley first encountered the principles of development and differentiation in Kingsley's *Water-Babies*. Having captured his interest in fiction, those principles then became a fund for analogies and concepts on which he drew to organize his adult work, producing a focus on development and differentiation as issues important in science as well. Fiction provided an analogy for Huxley's later scientific work in the case of the axolotl experiment, helping to structure the scientific questions Huxley asked about the world, and thus indirectly to determine the answers he found.

The axolotl experiment also illustrates the domaining effect: the subtle shift that takes place in ideas when they move from one cultural or social context to another. As the ideas moved from Kingsley's fairy story to Huxley's adult scientific work, they continued to reflect Kingsley's interest in development and differentiation. But—and here's the domaining effect—reflecting the new instrumental preoccupation of the scientific realm, Huxley did more than observe development and differentiation. He tried to reconstruct it. Moreover, as he transferred those principles—embodied by the water-babies and Mother Carey— from fiction to fact, what got lost was Kingsley's warning against meddling in nature's secrets. A fictional affirmation that there are limits to human knowledge became a scientific assertion that there should be no such limits.

The domaining effect means that cultural shifts in ways of thinking —as here, about reproduction—often occur in staggered and even contradictory fashion, on many discursive levels. For example, Huxley describes his victory in creating "a land animal which had not existed . . . for many thousands of years." This phrasing recalls Mother Carey's description, in *The Water-Babies*, of Prometheus, who could "turn the whole world upside down with his prophecies and his theories" (205). But Huxley's homage to Kingsley's fiction is undercut by the scientific work he is describing, which goes against the central theme of *The Water-Babies*.

Finally, the axolotl experiment story dramatizes how popular science writing, like science itself, is shaped by its broader cultural context. To the primary mission of popular science writing—explaining science to the layperson—it suggests we should add two secondary missions: merchandising and scientific public relations (or, to put it more bluntly, image doctoring). This moment of origin for Huxley's popular science writing career anticipates such contemporary phenomena as the exploding market for scientifically oriented, popular self-help books. But more than that, it anticipates the three essential features of our contemporary reproductive technologies: commodification of bodily control, scientific self-promotion, and the creation of a public informed about, and desirous of, such products. These three features would be explored at greater length just seven years after the axolotl experiment, in Huxley's short story, "The Tissue-Culture King."

SCIENCE, FICTION, AND TECHNOLOGIES OF REPRODUCTION

In "The Tissue-Culture King" Huxley focused on four reproductive technologies: endocrine treatment, experimental embryology, artificial parthenogenesis, and (its titular concern) tissue culture. Without undertaking an exhaustive or complete discussion of these extremely complex and diverse scientific fields, we can establish what these four technologies were most likely to have meant, in his day, to Huxley and to his readers.

Endocrine treatment is the administration of hormones to trigger developmental processes or in other ways alter the biological makeup of an individual. In Julian Huxley's day, endocrinology was appealing because of its potential to establish and police the boundaries of normalcy in both developmental and sexual behavior.[25] Scientists theorized that hormone imbalances could produce—and perhaps hormone treatments could therefore rectify—abnormal sexual development.[26] They also turned to endocrine treatment to avert or reverse the aging process. For example, Eugen Steinach and Serge Voronoff were experimenting with the grafting or injection of glands from other species into human beings.[27] Remembered now as the notorious "monkey gland" treatments, these ancestors of our contemporary estrogen-replacement therapy produced dramatic, if short-lived, rejuvenation effects in experimental subjects. As Julian Huxley explained in his essay "Sex Biology and Sex Psychology," they revealed that the "chemical directorate of the body [was] the interlocking system of endocrine glands" (139).

This new technique of endocrine-based rejuvenation would lead, Julian Huxley envisioned in 1922, to a new approach to the human body. In his essay "Searching for the Elixir of Life," Huxley predicted the growth of a system for human rejuvenation. Like cars, bodies will be repaired, in settings more reminiscent of a garage than a hospital:

> . . . great institutions for graft operations—human repair-shops. Men will have found methods for keeping organs alive outside the body, or they will be able to make grafts from tissue-cultures. Thyroids, pituitaries, adrenals, pineals, interstitial tissue, and many other regulating organs now unknown, will be in their several places, and aging humanity will come in to have their bodily system reanimated as cars come in to a garage to be overhauled.[28]

As Huxley's discussion indicates, the science of endocrinology was shaped (and its development anticipated) in terms set by its broader social context. We can see in his prediction the influence both of the instrumental, mechanistic approach of Fordist modern industry, with its organization of the body to meet the demands of the assembly line, and the increasing distribution and commodification of body parts and bodies in the postmodern era, which has produced our booming contemporary market for human organs and gametes and the thriving organ transplant business.[29] But endocrinology tugged the imagination backward as well as forward, from the world of science and technology to the world of literature and art. "Endocrinology has been called a modern form of mythology," enthused *Lancet* in a 1923 survey of endocrinological experiments.[30] Huxley's "Tissue-Culture King" would extrapolate its striking plot from the mixture of science and primitive religion conveyed in the assessment in *Lancet*.

Experimental embryology, like endocrine treatment, is concerned with the process of transformation. Huxley defined this field and summarized its achievements in "The Tadpole," an essay published in the same year as "The Tissue-Culture King." As he describes it, the field explores a question he first encountered in Kingsley's *Water-Babies*: "Does not each of us, in coming into this world, go through a transformation just as wonderful as that of a sea-egg, or a butterfly?" (55). In short, it is concerned with the development of life, from conception until birth. Experimental embryology got its start, Huxley explains, in the last decade of the nineteenth century, when zoologists—tiring of descriptive applications of evolutionary theory—followed Wilhelm

Roux, Hans Driesch, T. H. Morgan, and E. B. Wilson (among others) into the study of the physiology of development and the processes of differentiation.[31] Once again, Huxley's adult scientific writings return to issues first embodied in Kingsley's fairy story, in which the powerful image of Mother Carey embodies the principle of developmental differentiation. And just as Kingsley's Mother Carey resembles (and is at times replaced by) a white marble lady on a white marble throne very like Queen Victoria, so the field of experimental embryology is shadowed by imperialist constructions of its project. Metaphoric constructions of the field are remarkably similar from Kingsley's day to our own. They continue to attach imperialist and racist codings to its central concern: the idea of the process by which individual development is, or can be, controlled. To Kingsley's personification of the principle as Mother Carey/Queen Victoria, contemporary popular science has responded with its characterization of the contemporary understanding of the principle of embryological differentiation, the Hox genes, as the "potentates of animal development.[32]

Artificial parthenogenesis is the technique of activating an egg to produce cell division without contact with sperm, by shaking it or subjecting it to chemical or physical stimulus.[33] Although when Huxley wrote a survey of techniques for "The Determination of Sex," in 1926, "*artificial parthenogenesis* [had] so far only been tried upon animals which lay their eggs into the water before fertilisation," he predicted confidently that "it is theoretically possible in other forms, and that it would be only a matter of surmounting technical difficulties . . . to apply it to mammals and to human beings.[34] As Huxley describes it, this technique has implications that extend beyond the biological to the social:

> In many creatures . . . it has been found possible to make the egg develop without sperm. In sea-urchins the best method is immersion in certain chemicals; in starfish it is heat or shaking; in frogs it is pricking with an extremely fine glass needle which has been dipped in blood. The result is the same—that fatherless individuals are produced by man's intervention. (42)

Through this novel technique for reproduction, the father-progenitor is replaced by the male scientist, and "fatherless individuals" are produced by "man's intervention." As Huxley describes it, artificial parthenogenesis anticipates the scientific reconceptualization of fatherhood under the new reproductive technologies.[35]

Tissue culture is "a technique for maintaining fragments of animal or plant tissue or separated cells alive after their removal from the organism."[36] In 1922, in his essay titled (with bold romanticism) "Searching for the Elixir of Life," Julian Huxley enthusiastically predicted the far-reaching impact of experiments in the field of "tissue-culture." A passage from that essay contains the germ of the short story he would publish four years later, "The Tissue-Culture King":

> The strange facts of tissue-culture show that even in mammals most of the parts of the body are potentially immortal, and that it is only the system as a whole which is doomed to death. . . . Carrel in New York [has shown] . . . that fragments of living substance can be taken out of the body and grown in special culture media . . . [and] they can continue not only to live, but to grow for an apparently indefinite time. . . . *If this procedure had been known to primitive man, we should perhaps have found some nations seeking to keep their dead from corruption not by mummification, but by tissue-culture.* (625; my emphasis)

The research to which Huxley refers here was conducted by Dr. Alexis Carrel, at the Rockefeller Institute. This research had achieved a renown that spread beyond the scientific community to the broader culture, in part because of the promotion given it by the new mass-market genre of pulp science fiction.[37] As the editors of *Amazing Stories* explained in the sidebar to a piece published just four months before the "Tissue-Culture King": "The famous surgeon, Dr. Alexis Carrel, for the last fifteen years, has retained a fragment of a chicken's heart in a special medium, in which it has not only stayed alive and pulsating but has continued growing also, so that it only needs to be trimmed every little while, to be kept alive."[38]

Several kinds of crossover, or discursive drift, occur in the passage from "Searching for the Elixir of Life" in which Huxley predicts the future uses, and social implications, of culturing tissue: his analysis moves from fiction to fact, from Western technological culture to so-called primitive religious culture, from animal to human. Because they reappear in his short story, these transpositions are worth discussing in more detail. The tissues cultured by Carrel were not human, but avian. But when Huxley addressed the implications of tissue culture, he characteristically worked by analogy, shifting to consider the implications of the tissue-culture process for the human being. He also shifted the context for considering tissue culture from science to fiction in the course of his essay, mirroring a process of discursive drift common in the culture at large.

We can move from Huxley's treatment of the four reproductive technologies to other contributions to the same magazine in which his own short story was published: "The Ultra-Elixir of Youth" and "The Machine Man of Ardathia."[39] These other contributions further illustrate the movement of ideas about human reproduction and development between the realms of literature and science and their change —their domaining, if you will—in response to those different realms. These stories are worth pausing over, for they express concerns that reappear, with a slightly different emphasis because of the African setting, in Huxley's story: preoccupation with [the loss of] male sexual potency; scientific application of the technologies of modern industrial production; scientific focus on the process of aging; scientific interest in species evolution, devolution, and regeneration; and concern over the boundaries of the self. Illustrating the traffic between science and literature, these stories suggest the extent of Huxley's debt to his cultural milieu, as well as his culture's debt to his scientific work.

Billing itself as "a picturesque tale of the biological possibilities in the field of modern science," A. Hyatt Verrill's story "The Ultra-Elixir of Youth" takes as its subject the discovery of "a gas which will prove the Elixir of Life" (477). It fictionalizes the process of developmental reversal through endocrine treatment that Huxley discussed in "Searching for the Elixir of Life." In both, laboratory intervention produces developmental regression: in Huxley's essay, it takes place in the *Clavellina*; in Verrill's short story, it happens to adult male and female scientists (622). The story yokes embryos (babies) with test tubes (bottles), recalling the exchange of letters about babies in bottles that marks Huxley's first encounter with the intermingling of literature and science. The discovery of a new element, Juvenum, triggers a regression to embryonic existence reminiscent of Tom's experience in *The Water-Babies*. Yet this "Elixir of Life" is anything *but* the life-enhancing discovery dreamed of by the popular press covering Huxley's axolotl experiment. Rather than the joys of endless vigor, it brings the incomprehensible experience of "youth reduced to the nth degree, the utter youth of invisible existence, the youth of the prenatal, inexplicable germ of life . . ." (484).

"The Machine Man of Ardathia," by Francis Flagg (George Henry Weiss), is a tale of evolution, embryology, and extrauterine gestation or ectogenesis.[40] The protagonist is visited by a creature from the far distant future, more than 28,000 years hence. Enclosed in a cylinder of glass some five feet high, the creature not only recalls Kingsley's babies in bottles, but it anticipates the incubator used in in vitro fertilization in the mid 1980s.[41] The Machine Man seems to be gestating in an artificial uterus (799). (Ills. 8 and 9.) He tells the protagonist an evolutionary history of the human future, in which changes in reproductive method play the central role. From the prehistoric days when humans "reproduced their young in the animal-like fashion," to the era of the "Bi-Chanics," whose children were "introduced into incubators as ova taken from female bodies . . . brought to the point of birth in ectogenetic incubators . . . [and] had perfected the use of mechanical hearts," to the age of the "Tri-Namics," who "made envelopes—cylinders—in which they attempted to bring embryos to birth and to rear children," the human race finally came to the age of the Ardathians, the visitor and his contemporaries, who reproduce by a form of mechanical parthenogenesis (801–802). As the Ardathian explains: "The cell from which we are to develop is created synthetically. It is fertilized by means of a ray and then put into a cylinder. . . . As the embryo develops, the various tubes and mechanical devices are introduced into the body by our mechanics and become an integral part of it" (802).

As they co-narrate the history of human progress, the two characters in Verrill's story debate whether the notion of human evolution extends to the development from organism to machine. When the protagonist objects that Ardathians cannot claim human descent if they are "synthetically evolved and machine made," the visitor counters by pointing out the cyborgizing tendencies in 1920s biomedical science: "Did you not tell me you had wise ones among you who are experimenting with mechanical hearts and ectogenetic incubators?" The protagonist responds by alluding to Alexis Carrel's work at the Rockefeller Institute: "I have heard tell of chicken hearts being kept alive in special containers which protect them from their normal environment" (802).[42]

"The Machine Man of Ardathia," like many other works of fiction published in *Amazing Stories*, testifies to the interest in human origins, development, and the boundaries of the human lifespan and species current in the 1920s. These interests were expressed in plots about artificial gestational environments, synthetic growth hormones, and cyborg beings, as well as in allusions to actual scientific experiments, like Carrel's work with the chicken heart. (Ill. 10.) And these interests were expressed in images, too. The illustrations in *Amazing Stories* reveal uncanny parallels between the figuration of human conception and development in the early twentieth century and our current representation of reproductive technology. "The Ultra-Elixir of Youth" shows a

8 The "Machine Man of Ardathia"

baby surrounded by bottles (laboratory glassware, retorts, vials, and beakers) who is holding a test tube up to his scrutinizing eye. (Ill. 11.) And the illustration for "The Machine Man of Ardathia" not only anticipates the machinery for embryo culture as part of in vitro fertilization, but is an uncanny anticipation of our dominant current representation of the product of IVF—the test-tube baby—in the sketch of

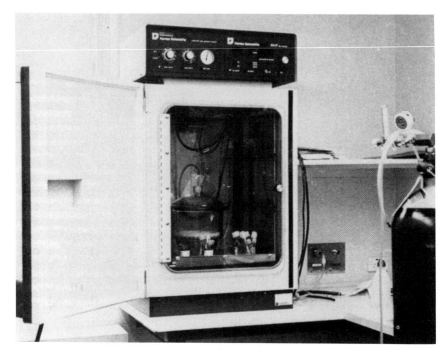

9 A Contemporary Incubator Used in *In Vitro* Fertilization

a dome-headed, naked creature enclosed in a transparent oblong glass tube.

This survey of how the four focal reproductive technologies were popularly understood in the 1920s reveals that certain ideas appear in and travel between the scientific and fictional domains: (1) an analogy between animal and human reproduction, development, degeneration, and mutation; (2) an ambition to use these technologies to manage, control, and police the boundaries of sex, race, and species; and (3) a tendency (driven by eugenic visions of human perfection) to conceptualize the human being as a machine. Huxley's "Tissue-Culture King" suggests how reproductive technologies can embody both the appeal and the danger of these ideas in a global arena.

"THE TISSUE-CULTURE KING"

Appearing first in the *Yale Review* cheek-by-jowl with E. M. Forster's review "The Novels of Virginia Woolf," "The Tissue-Culture King"

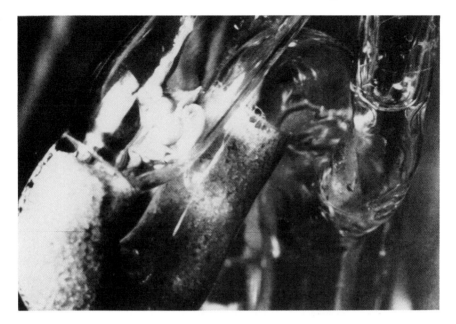

10 Alexis Carrel's Tissue-Cultured Heart

was quickly reissued in *Amazing Stories*, the SF magazine launched in the United States just the preceding year.[43] We might describe the story as a generic chimera—a combination of several different species of story—for its plot lays claim to roots in both canonical high culture and popular culture, reflecting its ambiguous and multiple social locations. As Huxley explains its origins in his *Memories*, the tale was inspired by tales of king worship in West Africa, especially Frazer's *Golden Bough*, and it merged the sort of African adventure story found in boys' weeklies with a science fiction fantasy (219). The mix was perfect for the newly founded *Amazing Stories*, whose editor, Hugo Gernsback, "hoped to teach science through fiction as well as to cultivate an interest in the potential of technology."[44] Yet breaking from both traditions, Huxley's story concludes with a moral attacking the popular press and exhorting a high-minded reevaluation of the scientific project.

The process of analogy provides both plot and central metaphor for "The Tissue-Culture King." It links science and religion, "first world" and "third world," normal and abnormal, human and animal, self and other. Yet in the very process of figuring the parallels between these

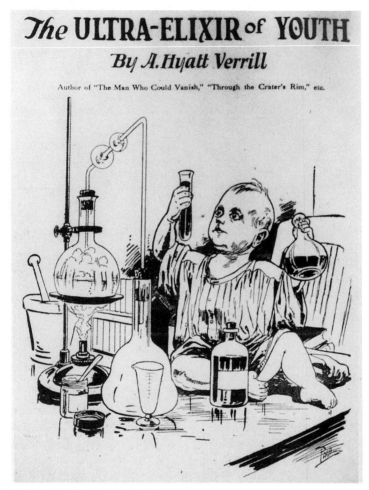

11 The "Ultra-Elixir of Youth"

realms, analogy also introduces the possibility of their differentiation, and that tension between like and unlike drives Huxley's story. The plot has a Eurocentric, racist framework: a white explorer in Africa is taken captive by "savages" and brought to the tribal headquarters. There he meets another captive white man who, in order to save his life, has invented a role for himself within the tribal order: a role he finds so satisfying that he declines the newcomer's offer of rescue, preferring to stay with the life he has made. A biological cast to this familiar tale distinguishes it both from high-modernist critiques of imperialist exploration and the anthropological tales popular at the time. Like Virginia Woolf's *Voyage Out*, praised by E. M. Forster in the neighboring essay in the *Yale Review* as "a strange, tragic, inspired book whose scene is a South America not found on any map and reached by a boat which would not float on any sea, an America whose spiritual boundaries touch Xanadu and Atlantis," the short story takes us to an Africa that is also fantasy to the core.[45] But the fantasy is not social, as is Woolf's, but biological.

Whereas Conrad's Marlow must journey to the heart of darkness through a country whose social deformations (heads on posts outside native villages) hint at a social horror to come, the experiences of Huxley's white explorer foreshadow an impending horror that is specifically biological. Ironically inverting the trope of the African captivity narrative, Huxley has his explorer-protagonist brought to an African capital city remarkable not for its primitive squalor, but for its advanced technology.[46] The Kurtzian white man whom the protagonist meets there in the jungle is not an imperialist exploiter of African raw materials (a hunter or a trading company agent) but rather a scientific exploiter of African experimental subjects (a British medical researcher, bringing "medical aid" to the Africans). "Hascombe, lately research worker at Middlesex Hospital, now religious adviser to His Majesty King Mgobe," originally came to Africa to study sleeping sickness (482–483).

Captured by the African tribe, Hascombe is charged with intruding into their secret "holy place" and threatened with execution. Yet Hascombe has intellectual resources that save his life. He "had interested himself in a dilettante way in anthropology as in most other subjects of scientific inquiry," from which we can conclude that he has acquired a facility in perceiving and working with analogy.[47] Having observed that "one of their rites was connected with blood," Hascombe draws on that analogical facility, and his medical training, to save himself. "You revere the Blood," he tells them through the interpreter. "So do we white men; but we do more—we can render visible the blood's hidden nature and magic" (484–485). Showing them a drop of his own blood through the lens of his "precious microscope," Hascombe explains to the natives "that the blood was composed of little people of various sorts, each with their own lives, and that to spy upon them thus gave us new powers over them" (485).

Hascombe determines that the main motifs of the tribal religion are analogous to four Western scientific procedures—"Tissue culture; experimental embryology; endocrine treatment; artificial parthenogenesis"

—and he determines to use them to produce as much power as he can (487). By linking religion to science, and using both to consolidate state power, Hascome might be said to have founded the first *pharmacracy*: the term comes "from the Greek root *pharmakon* for 'medicine' . . . [and is] analogous to 'theocracy,' rule by God or priests, and 'democracy,' rule by the people."[48] Moreover, in his deployment of the four reproductive technologies to produce power, he anticipates the later site for which the label "pharmacracy" was coined: the contemporary political and social power enjoyed by biomedicine, which has given rise to a growing for-profit reproductive technology industry.

Centerpiece of Hascombe's pharmacracy is his "Institute of Religious Tissue Culture," which operates on the body of the king itself in a uniquely Western, instrumental deformation of the West African king-worship about which Huxley had recently read in the *Golden Bough*, with its discussion of the central role in primitive societies of the "sacred priest-king" (488). The Institute's project originates in ideas transferred from one context to another: from British science to African religion. As Hascombe tells it: "The most important new idea which I was able to introduce was *mass-production*. Our aim was to multiply the King's tissues indefinitely, to ensure that some of their protecting power should reside everywhere in the country" (490). The king's sacred tissues are surgically excised, grown in a culture medium, partitioned, and distributed to his subjects. His identity assaulted, extended, divided, proliferated, and changed, he becomes "the tissue-culture King."

Hascombe's institute parodies the forces of modern, commercial medicine in its emphasis on lifestyle reeducation, consumer-orientation, and sex-linked marketing. The king's tissues are commodified: "To everyone subscribing a cow or a buffalo, or its equivalent . . . a portion of the royal anatomy should be given, handsomely mounted in an ebony holder" (491). In addition, following the best marketing techniques, Hascombe makes sure that "tissue-culture renewal" and subculturing rights are available by annual subscription. And finally, workers to carry out the routine task of tissue-culturing are recruited by offering to give them a "permanent [sex-linked] status" to their position: "Sisters of the Sacred Tissue." "From this, with age, experience, and merit, they could expect promotion to the rank of mothers, grandmothers, great-grandmothers, and grand ancestresses of the same" (491).

In its gender hierarchy and its antiseptic technological aura, Hascombe's Institute of Religious Tissue-Culture recalls the Rockefeller Institute laboratory of Peyton Rous, which Huxley visited shortly before World War I. Rous had formulated a technique for tissue culture that would allow him to "grow fragments of chick embryos," and Huxley was "excited by the new method of grafting tissues on to the egg's surface membrane" (*Memories*, 85–86). In its gender hierarchy and its antiseptic technological aura, Hascombe's institute also anticipates the Rockefeller Institute laboratory of Alexis Carrel, which Huxley would visit in the interwar period. As he describes it in *Memories*, it had "room after room full of the latest apparatus, attended by rows of female assistants and secretaries all garbed in white, and mostly very pretty" (168). Hascombe's visitor decides that "the Hascombe Institute was, it is true, not so well equipped, but it had an even larger, if differently colored, personnel" (488). Huxley's scientific parable makes its point by reversing the modernist tale: the impact of Hascombe's Institute lies not in its divergence from Western science, but in its ironic, exaggerated, parodic resemblance to it. Like Alexis Carrel's laboratory, Hascombe's Institute relies on the enabling existence of gender hierarchies in the scientific workplace—hierarchies that were solidified with the institutionalization of modern science, as Londa Schiebinger has shown—even as it breaches the boundaries of the body, of species, and of life to produce political power.[49]

As in Western biomedicine, the revenues from the commodification of the king's tissues are poured into research, which diversifies first into the "ancestry-worship" branch of religion, and then out of tissue-culture into endocrinology, experimental embryology, and parthenogenesis. The result is a fantastic pair of additional institutes which produce power by manipulating not the king's body, but the bodies of the tribe's subject populations: animals and women. Here Julian Huxley, like his brother Aldous and their close friend Naomi Mitchison, anticipated the facilitating relationship between animal husbandry and human reproductive medicine, based on the operation of analogy.[50]

"Home of the Living Fetishes" extends the king's domain to the animal kingdom, producing power for this African pharmacracy in a process analogous to that used by Western science: the laboratory-based generation of biological anomalies, like the axolotl metamorphosis. Hascombe's description of this institute is a surreal mixture of industrial and scientific rationality. Central is the implicit analogy between the aesthetic monstrosity and the genetic mutation:

> I thought I would see whether art could not improve upon nature,
> and set myself to recall my experimental embryology. . . . I utilize the

plasticity of the earliest stages to give double-headed and cyclopean monsters. That was, of course, done years ago in newts by Spemann and fish by Stockard; and I have merely applied the mass-production methods of Mr. Ford to their results. But my specialties are three-headed snakes, and toads with an extra heaven-pointing head. (495)[51] (Ill. 12.)

Hascombe's "Factory of Ministers to the Shrines" is devoted to "endocrine products." Explaining this institute to his visitor, Hascombe recalls the conglomeration of scientific experimentation, mass-marketed patent medicines, and popular science and self-help writings that sprang up in the 1920s around hormone treatments.[52] Like Huxley with his axolotl, this institute uses hormone injections to produce deliberate freaks, but here the experimental subjects are not Mexican amphibians, but African human beings. The result: bio-factory-built objects of worship. The dwarfs, half-wits, "adolescent girls . . . with the most copious mustaches [who] found ready employment as prophetesses," tremendously fat women, and "children sexually mature at seven or eight" produced in this "Factory" join the products of the "Home of the Living Fetishes" to satisfy the nation's desire to worship something Other, and thus contain the population's transgressive impulses (494).

Hascombe's final, still-unfinished experiment consolidates the impulse common to all the zoological, embryological, and eugenic programs of his institutes: the appropriation and control of reproductive power. Guided by the analogy between animal reproduction and human reproduction, Hascombe plans to "apply Jacques Loeb's great discovery of artificial parthenogenesis to man, or, to be precise, to these young ladies," and so to "grow a race of vestals, self-reproducing yet ever virgin" (494–495). The reference is to a classic experiment performed by Jacques Loeb in 1901, which demonstrated that unfertilized sea-urchin ova would develop to the larval stage when mechanically stimulated.[53] Loeb's experiment marks an important shift in the nature of scientific work between two different orders of knowing. Loeb moves from recognition to construction, as Strathern frames them in the second epigraph to this chapter. Or, as Robert L. Duffus explained in the *Century Illustrated Monthly Magazine*, "Loeb . . . altered the role of the scientist from that of the mere observer to that of an engineer and even creator. If science follows the highway marked out by these men [such as Loeb], mankind will be at last not the servant, but the master of its environment."[54]

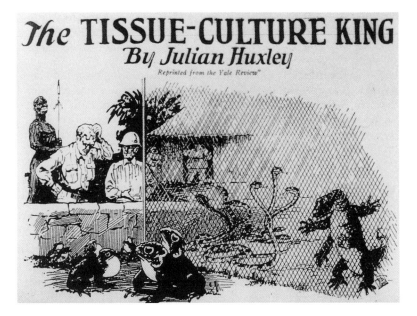

12 The "Home of the Living Fetishes"

"The Tissue-Culture King" ends with an escape attempt, foiled not by the physical force of the tribesmen, but by mental connections between them and their English captives, links forged in the final institute, which was dedicated to work on "reinforced telepathy" (496). This final plot element turns the tables on the analogizing facility that has been central to all of Hascombe's previous accomplishments. His achievements in all the other institutes were made possible by his focus on likeness rather than difference: of the king to his tissues; of the African religions to Western science; of toads and snakes to humans; of blood-based initiation rites to Western microscopic medicine. But here, it is precisely the likeness between the tribesmen and Hascombe that makes his escape impossible, a likeness enforced by the institute's development of mass-communication and control strategies echoing Western propaganda. "Tormented by doubts as to whether the knowledge of mass-telepathy would not be a curse rather than a blessing to mankind," the explorer who escapes declines to "bring Hascombe's discoveries before the Royal Society or the Metaphysical Institute" (504). Instead, he closes the story with a warning, a "sermonical turn," that seems to swerve from science to religion:

The question I want to raise is this: Dr. Hascombe attained to an unsurpassed power in a number of the applications of science—but *to
what end did all this power serve?* It is the merest cant and twaddle to go
on asserting, as most of our press and people continue to do, that increase of scientific knowledge and power must in itself be good. I commend to the great public the obvious moral of my story and ask them
to think what they propose to do with the power which is gradually being accumulated for them by the labors of those who labor because
they like power, or because they want to find the truth about how
things work. (504)

THE POWERS AND DANGERS OF ANALOGY

What is the function of analogy as a scientific practice? What are its
effects, its powers, and its dangers? And finally, what does analogy
have to do with the development and practice of reproductive technology? As "The Tissue-Culture King" reveals, analogy not only helps us
to extend scientific practices and thus further knowledge, but it also
encourages the processes of objectification (putting people in the position of things) and totalization (applying broad principles in disregard
of specific local practices). Finally, analogy has a constructive rather
than a descriptive relation to the field of reproduction, as to the other
scientific fields in which it operates.

Clearly, analogy is a powerful device in "The Tissue-Culture King."
It is his ability to draw analogies that saves Hascombe's life. His realization that the tribal religion is analogous to Western scientific practices ("You revere the blood. So do we white men . . .") prompts him
to apply Western scientific/medical techniques (themselves developed
through the process of analogy) to new subjects, once again following
the principle of analogy (if a chicken's tissues can be cultured, so can
a king's). Analogy provides the principle tenet of Hascombe's new
Institute of Religious Tissue Culture: that a bit of the king's tissue is
analogous to the king itself, thus worthy of worship, the tithing of livestock, and the service of women. And even the Institute's assembly-line
operations implicitly reflect an analogy between the industrialized mass-
production of the king's tissues and the reproduction of the king's
lineage, making it possible for Hascombe to increase the efficiency of
the tissue-culturing process, and thus rapidly to magnify his tribal
power. Still other analogies, anticipating contemporary biomedicine,
provide him with access to biological and spiritual control. Based on an
audacious analogy between "a necropolis [and] a histopolis," "a ceme-
tery [and] . . . a place of eternal growth," Hascombe's "repository of

the national tissues," like our contemporary Human Genome Initiative, permits unprecedented control of the nation's genetic materials.[55] The hormone treatments carried on at the "Factory of Ministers to the Shrines," anticipating our contemporary fascination with the steroid-enhanced prowess of athletes, produce monstrous anomalies for the population to worship (493).

Analogy also has its dangers, dramatized by the experiment in telepathy with which the story ends. Hascombe realizes that the native tribesmen—like the Europeans—possess a remarkable receptivity to hypnosis. An attempt to hypnotize the natives recalls "the most startling cases of collective hypnosis recorded by . . . French scientists" (497). The technique of "super-consciousness" that Hascombe induces in the natives transmits itself to him too. With his flight nearly completed, he is unable to resist the suggestion that he return to the village: "He must go back; he knew it; he saw it clearly; it was his sacred duty" (503). As Hascombe discovers, analogy works both ways: likeness (including the suggestion of mental likeness) produces control that eventually controls the controller.

ANALOGY AND THE DOMAINING EFFECT

Analogy not only fueled the development of Hascombe's fictional scientific institutes in Huxley's short story; arguably, it also fueled the development of a scientific field. Five years after Huxley published his short story, his brother Aldous Huxley published *Brave New World*, a dystopian novel inspired by the writings and scientific findings of Julian Huxley, his friends, and colleagues.[56] Within the next six years, John Rock and Arthur Hertig would lay the foundations of in vitro fertilization with a series of experiments on fertilized human ova. And Rock and Hertig acknowledged that they were "partly inspired in their reproductive research by a novel which had been published a few years earlier, Aldous Huxley's *Brave New World*."[57]

The dystopian vision of reproductive technology embodied so forcefully in both Julian and Aldous Huxley's texts did not travel along with the technology, however. No sense of danger clouds the optimistic vision of Rock and Hertig's intellectual progeny, reproductive scientists Robert Edwards and Patrick Steptoe. In response to the challenges of ethicists and theologians that in vitro fertilization could produce unacceptable results—"Suppose the baby born is abnormal, a cyclops say, or some other monster?"—Edwards and Steptoe responded, "Most of the time . . . our preoccupations have not been concerned with ethics; rather they have been dense with scientific speculations, technical problems" (*Matter of Life*, 4–5). In place of Huxley's anxious question ("To what end?"), we have a shift of focus from the result to the intention, as Edwards and Steptoe declare, "Our aspirations were only to help" (4).

In its image of African peoples subjected to the operations of reproductive technology and changed, even mutated, in order to produce power for others, the plot of Huxley's short story suggests the dangers of analogical thinking. If animals can be used as experimental subjects to produce power through the manipulation of their bodies, analogy suggests that the same is true of human beings. If the reproduction of animal populations can be controlled and manipulated to produce power, by analogy the reproduction of human populations can also be subjected to control and manipulation. In each case, the analogy makes connections without drawing attention to differences: the difference between an individual animal and an individual human being; the difference between manipulating animal populations and exercising that kind of dominance on human populations. The "unsurpassed power" Hascombe produces follows directly from the application of the animal-human analogy.

Huxley's story is not the first to show a scientist producing unsurpassed and unacceptable power in that particular way. From Mary Shelley's Victor Frankenstein to H. G. Wells's Dr. Moreau, writers have given us cautionary examples of the dangerous consequences that could follow from the animal-human (as well as the human-machine) analogy.[58] What is unusual about Huxley's story is its premise: that Hascombe came to Africa to undertake a medically progressive intervention, an attempt to improve the health of the people by discovering the vectors transmitting sleeping sickness. In short, Hascombe's task in Africa was precisely to use the analogy between animal and human to minister to a population. His power results not from the misuse or misinterpretation of scientific practice, but from its extension to a global field.

EUGENICS AND ANALOGY

If we move back from a close focus on the text of "The Tissue-Culture King" to consider its context, we find that eugenics is the horizon of Huxley's story: Hascombe's emphasis on medical intervention to improve the health of a population embodies the mission of that new, parascientific field. As defined by its founder, Sir Francis Galton, "Eugenics is the science which deals with all influences which improve the inborn qualities of a race; also with those which develop them to the utmost advantage." A population-management program combining benefi-cent medical attention to reproduction with more proscriptive inter-ventions into individual and group reproductive behavior, as Daniel Kevles has observed, mainline eugenics was fueled as much by a fear of the racial Other and racial degeneration, as by the desire to improve the human lot.[59] Eugenics functioned analogically, applying insights gleaned from work in animal husbandry and animal genetics to the project of improving the stock of the human species.

Huxley was a crucial member of the reform wing of the eugenics movement, detailing the abuses of reproductive medicine that sur-faced in the years immediately preceding World War II and calling in contrast for sound scientific practice. Later, as Donna Haraway has shown, Huxley reflected postwar liberalism in advocating at most a sci-entifically informed, environmentally calibrated, and demograph-ically framed manipulation of the human species.[60] But his susceptibility to the powers and dangers of analogizing led to a contin-uing conflict in his relations with eugenics. In the years following his publication of "The Tissue-Culture King," he would, paradoxically, celebrate some medical interventions into the reproductive activities of individuals and populations, even while continuing to critique eu-genics for its susceptibility to bias and misuse. For example, in his 1936 Galton lecture "Eugenics and Society," while castigating eugenicists for irresponsible scientific practices, he affirmed that a reconstructed eugenics would be the "religion of the future"—a bizarrely straight-faced reprise of his humorous premise for "The Tissue-Culture King."[61]

The story of Huxley's response to one specific eugenic technique—*eutelegenesis*, or artificial insemination for eugenic purposes—rounds out my consideration of the powers and dangers of analogy. In "Eu-genics and Society," Huxley singled out eutelegenesis for praise as the last stage in a welcome process of reproductive evolution.[62] Separat-ing love from reproduction, eutelegenesis "[makes] it possible to pro-vide different objects for the two functions," Huxley argued. "It is now open to man and woman to consummate the sexual function with those they love, but to fulfill the reproductive function with those whom on perhaps quite other grounds they admire" (78).

Eutelegenesis was a new name for an old technique, developed in animals and extended to humans during the course of the eighteenth and nineteenth centuries, whose widespread possible human applica-tions were given enthusiastic publicity by eugenicist Herbert Brewer in

England and geneticist Hermann Muller in the United States. Eutelegenesis originated in 1779, with Lazzaro Spallanzani's artificial insemination of a spaniel.[63] The technique was first applied to human infertility by Dr. John Hunter, who earlier improvised the technique to address the congenital penile deformity of a linen draper which prevented him from fathering children.[64] More recently, according to Brewer, who traced its history in an article in the *Eugenics Review* in 1935, it had been used with tremendous success by the Moscow Experimental Station, whose results up to 1932 included "two million cows, three million ewes, 650,000 mares and 200,000 sows . . . successfully impregnated by artificial means."[65] Recalling primitive peoples' conviction that conception occurred independent of sexual intercourse, Brewer claimed that with eutelegenesis, "science has transformed primitive fiction into modern reality" (122).

Eutelegenesis was propounded in the United States by Hermann Muller, an American geneticist whom Huxley had recruited to join him at Rice University in 1914, where he helped to found the biology department. Muller's enthusiasm for eutelegenesis had been inspired by the endocrinological work of Gregory Pincus at Harvard, who in 1934 mastered in vitro fertilization in monkeys and by 1939 had also accomplished surrogacy in rabbits, as well as by the news in 1935 that Alexis Carrel had managed to keep mammalian ovaries alive and growing in tissue culture outside the body.[66] Like Brewer, Muller wondered if a technique that worked in animals would, by analogy, work in human beings, and he began to map the scientific developments that would be necessary for a true eugenics of human reproduction. Muller and other socialist biologists of his era planned to use scientific techniques to reshape the human population and so produce a more equitable distribution of resources. As Kevles describes it, the plan they evolved for producing greater human harmony through reproductive evolution strikingly recapitulates the narrative of "The Machine Man of Ardathia," in its progression from eutelegenesis (artificial insemination) to "penectogenesis" (in vitro fertilization), to ectogenesis (extrauterine gestation) (*In the Name of Eugenics*, 188–192).

Like Huxley, Muller critiqued eugenics for its inattention to the social and environmental factors shaping human development, as well as for its susceptibility to bias. Indeed, his biographer reports that Muller "never permitted his name . . . to be used in connection with any eugenics movement, as each organization espousing eugenics turned out to be flawed by the fallacies of racism, unjustified elitism, class bias, misplaced emphasis, or oversimplified genetic determinism."[67] In his book *Out of the Night* (1935), Muller called for biological improvement *only* as a foundation for widespread and lasting socioeconomic and political reform, which would lead to "a more truly co-operative basis of society."[68] Observing that "eugenics . . . has become a hopelessly perverted movement," he accused it of doing "incalculable harm by lending a false appearance of scientific basis to advocates of race and class prejudice, defenders of vested interests of church and state, Fascists, Hitlerites and reactionaries generally" (*Out of the Night*, 10–11). Julian Huxley praised the eutelegenetic program of *Out of the Night* for its social vision, for which the first step was biological improvement of the human species.

Yet if Huxley explicitly stressed reproductive technology's power to increase human liberty and freedom of choice, while Muller affirmed biological improvement as the basis for a cooperative egalitarian society, their writings also betray a metaphoric subtext more concerned with the exercise of power than with the attainment of freedom. We have seen how Huxley's "Tissue-Culture King" ends with the question "to what end did all this power serve?" Muller's *Out of the Night*, too, is concerned with bio-power.[69] Yet it addresses that theme via an analogy that recalls not Huxley's cautionary question, but rather the medical imperialism of the rest of "The Tissue-Culture King" and the metaphor of global exploration in the passage from Huxley's "Science, Natural and Social" that forms the epigraph of this chapter.

I close with an examination of Muller's analogy in *Out of the Night*, as a way of returning to the question that opened this chapter—the powers and dangers of analogy, as it circulates between literary and scientific uses:

> Before we can properly understand the living things of our world we must first know the structure of that new world-of-the-small which within the past few decades has been opened to the mind of man. . . . [M]odern genetics is already beginning to invade the ultra-microscopic land inside the eggs and sperm cells . . . and to bring back from this survey what we actually call "maps," on which are shown . . . the locations of hundred of the separate hereditary particles or "genes" that help to determine the various visible characters of the individual growing from that germ cell. (22–23)

Despite his deliberately articulated, conscious intention to avoid race and class prejudice and the atrocities carried out by "Fascists, Hitlerites, and reactionaries generally" in the name of eugenics, Muller here uses an analogy whose implications are both familiar and troubling. The new science of genetics is like exploration, he observes; the interior of the body is analogous to an unknown territory; microscopic study is like mapping. And he concludes this survey of the frontiers of science with the assertion: "To gain adequate control over the world of things of our own size, then, we must first seek knowledge and control of the very small world. . . . [F]or man, the road to the macrocosm lies through the microcosm" (22–23).

How to read the implications of this analogy? One way would be to see it as exemplifying a shift in scientific discourse. Muller was an early and powerful advocate of the reorganization of the biological sciences, according to Evelyn Fox Keller. He held that the field had undergone three crucial changes in its knowledge base: The "relocation of the essence (or basis) of life" to the gene; the "redefinition of life" from the complex of qualities of living organisms to a set of instructions; and a "recasting of the goals of biological science" from a generalized sense of description or even broad-based intervention to a very focused intervention in order to control the very processes of "making and remaking life."[70] (Shades of Mother Carey, "making things make themselves.") Because he was an early advocate of these three shifts, Keller explains, Hermann Muller felt that biology needed to rethink its institutional goals and practices. By that reading, Muller's analogy exemplifies the redirected goals and practices called for by the three shifts in the structure of biological knowledge: a focus on genes rather than organisms; a stress on information storage and retrieval (in its image of maps); and, most explicitly, a stress on control rather than observation, particularly over gametes.

But this passage reflects more than just a shift in the culture of the biological sciences. It also reveals a cultural stability: the remarkably persistent representational strategy of drawing an analogy between scientific experimentation and global exploration. If its focus on genes, maps, and control shows us that things have changed dramatically in modern biology, that attention has been redirected from the organism to the molecule, this explorer analogy suggests that some things have stayed the same. It hearkens back to the age of imperialism rather than forward to the age of what Haraway has called informatics.[71]

Why the carryover of this old analogy into a new scientific era? This is the process of domaining in action, and feminist cultural studies of science can alert us to some of its implications. Evelyn Fox Keller has eloquently and extensively studied the gendered implications of the analogy central to Muller's passage: the search for scientific knowledge figured as the visual penetration of nature, imaged as a body. As she has shown, this analogy renders the object of scientific study (nature) a passive corpse/*corpus*: it equates scientific discovery with the act of unveiling or making visible, with its gendered suggestion of a male science ravishing female nature, and it is motivated by an unconscious equation between the (female) secrets of life and the (male) secrets of death.[72]

Yet for my purposes this passage is most interesting in its collapse of the body of nature into the nature of the body, and the opposition of that conjoint entity (nature/body) to a spatial expanse imaged as *terra incognita*. Here the gendered practices of science are racialized: the scientist becomes the European explorer mapping the dark continent, achieving control over all its little denizens, who are (gender now returning) figured as gametes subject to scientific control through the microscopic practices of reproductive power/knowledge. Let us return to Huxley's description of the uses of analogy in science, from the epigraph to this chapter: "Analogy is . . . the clue which guides the scientific explorer towards radically new discoveries, the light which serves as the first indication of a distant region habitable by thought." Here we see a racial subtext similar to Muller's "world of the small." A scientist using analogy, in short, is like an explorer using a light: they both must discover, penetrate, and then settle "their" territory. The object of scientific knowledge here, too, has been racialized: it is a non-Western country, subject to imperial conquest. Scientific knowledge practices are also racialized and sexualized: they unveil and colonize conceptual territory that is simultaneously a woman and a person of color.

How do we understand Huxley's and Muller's sincere, even vehement critiques of the racial and sexual oppressions springing from eugenics after hearing them describe their own scientific practices using analogies that figure the conjoined control of race and sex? As evidence, perhaps, of the complex tensions between emancipatory and controlling impulses in the work of these two representative twentieth-century socialist scientists.[73] But more broadly speaking, the contradiction also illuminates the operation of analogy in the development and representation of reproductive technology. It teaches us that if analogies move us forward in science (as Huxley argued), they also hold us back in society. Because of the domaining effect—the survival of naturalized ways of thinking when a set of ideas travels from one cultural setting to another—they preserve and thus perpetuate (albeit usually unconsciously) earlier cultural forms, which are enacted slightly differently in each new scientific area. Finally, the contrast between the emancipatory goals and oppressive metaphors of their writings suggests that the representations of a scientific field such as reproductive technology may illuminate both the boundary conditions for its existence as well as the impetus for its development.

When we understand that the origins of reproductive technology lie in an analogy between animal embryology and human embryology, our attention is drawn to two unexamined, persistent components of this newly constituted field. Objectification (taking a person as a thing) and totalization (generalizing from a principle in defiance of local data and practices) were essential operations if animal embryology was ever to be applied to human embryology. Essential, and yet taboo: thus the process of analogy functions both to make connections and to mask connections. If we refuse to accept Strathern's notion that analogies are simply "artificial graft(s) that [bear] no intrinsic relationship to the subject of embryo development" (or, in this case, the subject of reproductive technology), and if we instead interrogate the analogies foundational to reproductive technologies in their earliest years, we find embedded there deep conflicts over the two most problematic aspects of contemporary reproductive technology: the tendency to use people as things (shadowing a range of practices from surrogacy to embryo research) and the tendency to totalize (leading to the erroneous conclusion that any reproductive technology means the same thing in all social, economic, cultural contexts).

This interrogation of the role of analogy in the development of the field suggests a new direction for feminist analysis of the new reproductive technologies. First, we need to be alert to the tendency to objectification embedded in these technologies, with oppressive consequences for race, class, or even the species. The roots of this tendency lie in the process of analogizing reproduction and development from the animal to the human, and from the individual to the species, a process in which the new reproductive technologies originated and which possesses eugenic and evolutionary overtones. Second, we need to resist the tendency to make totalizing analyses of these new technologies: to assume that they mean the same thing (whether good *or* bad, in all sites, at all times, for all women). The roots of this tendency, too, lie in the act of analogizing in which the field was conceived, an act that as Huxley himself both observed and involuntarily illustrated, "may very readily mislead."[74]

two

The Ectogenesis Debate and the Cyborg

IMAGING THE PREGNANT BODY

> Scientific and popular conceptions of human embryology and ob-
> stetrics bring into focus issues relating to gender, the construction
> of identity, modes of creation and production, the mother-child
> relationship, and paradigms of nature and the natural.
> —ANDREA HENDERSON, "Doll-Machines and Butcher-Shop Meat"

In 1982, spurred by the birth four years earlier of the first "test-tube"
baby, the British government established the Warnock Committee of
Inquiry into Human Fertilisation and Embryology. The Committee's
brief was wide-ranging: to investigate "current and potential developments in medicine and science related to human fertilisation and embryology," from social, ethical and legal perspectives, and to recommend
appropriate legislation to the government.[1] Surveying such "techniques for the alleviation of infertility" as artificial insemination, in
vitro fertilization, egg and embryo donation, surrogacy, and egg and
embryo freezing and storage, in 1984 the Committee issued recommendations that were decisively affirmed six years later. In 1990, after
an extensive debate, the House of Commons voted 364 to 193 to ratify
the Human Fertilisation and Embryology Bill, whose central provision
was the licensing of experimentation on human embryos for a period
of up to fourteen days, or until the appearance of the so-called primitive streak.[2]

The implications of the Warnock Committee findings and the Human Fertilisation and Embryology Bill have received extensive analysis by a number of feminist theorists, and I will not attempt to
summarize them here.[3] Rather, I want to return to one specific recommendation of the Warnock Commission that has, to date, received
little attention. In the section headed "Possible Future Developments

in Research," the Warnock Committee recommends that a procedure known as "ectogenesis," or extrauterine gestation, be outlawed:

> 12.7. It has been suggested that in the long term further development of current techniques could result in the maintenance of developing embryos in an artificial environment (ectogenesis) for progressively longer periods with the ultimate aim of creating a child entirely *in vitro*. . . .
>
> 12.8. We appreciate why the possibility of such a technique arouses so much anxiety. There are however two points to make about this. First, such developments are well into the future, certainly beyond the time horizon within which this Inquiry feels it can predict. Secondly, our recommendation is that the growing of a human embryo *in vitro* beyond fourteen days should be a criminal offense. (Warnock Report, 71–72)

A fascinating paradox grounds this quiet, two paragraph recommendation: while asserting its unwillingness to engage in futurological speculations (which it describes as beyond its predictive competence), the Committee simultaneously engages in them, by recommending explicitly that a still only hypothetical reproductive technology be outlawed.

The context for this paradoxical intervention by the Warnock Committee is a contemporary discussion of ectogenesis that has focused primarily on two issues: the likelihood of its successful development and its social (as well as specifically feminist) implications. Philosopher Julian S. Murphy has summarized the steps necessary for the creation of an artificial womb:

> If ectogenesis is to be accomplished, replacements must be found for the series of biochemical processes performed by women's bodies in pregnancy: egg maturation, fertilization, implantation, and embryo maintenance, temperature control, waste removal and transport of blood, nourishment, oxygen to the embryo. Such a procedure, if successful, would accomplish *in vitro* gestation (IVG) for human reproduction.[4]

Opinion varies on whether such a sequence of procedures could be successfully completed. British embryologist Anne McLaren is doubtful:

> To adapt such technology from the mouse to a species such as our own, with a large fetus and a protracted gestation period, would be a colos-

sal task. At present there seems no reason to think it would ever be attempted; even if vast resources were to be devoted to it, the chances of achieving a successful result before the end of the twenty-first century would be slim.[5]

Feminist opinion is divided, with some critics taking the position that advances in neonatology make ectogenesis inevitable, and others arguing that the expense and technical complexity of the procedure make its development and deployment highly unlikely.[6] Yet with rare exceptions, the general attitude toward ectogenesis at the present time is negative, ranging from dismissive distaste to alarmed anticipation.[7]

The contemporary negative assessment of ectogenesis may have been one of the phenomena that the Warnock Committee's paradoxical recommendation was designed to address, by making a crucial distinction between embryo research (necessitating a brief period of in vitro embryo growth) and ectogenesis (a prolonged period of in vitro gestation). Although a significant portion of the popular press and parliamentary debate framed the bill (either positively or negatively) as legalizing and enabling embryo research, paragraphs 12.7 and 12.8 from the Warnock Committee report can be read as establishing the foundations for an alternative, more wide-ranging interpretation of the legislation, as outlawing and preventing extrauterine gestation of human embryos. One way of understanding this paradox is to see the Warnock Report as a maneuver to position the later legislative actions (the Human Fertilisation and Embryology Bill) in a reassuring light.

Yet this paradox might be more than a strategic maneuver: it may also express an unrealized desire. In his 1925 essay, "Negation," Freud suggests what that desire might be when he observes that "the subject-matter of a repressed image or thought can make its way into consciousness on condition that it is denied."[8] Invoking ectogenesis in order to deny (outlaw) it, the Warnock Commission may be seen as satisfying a desire: putting into circulation the very same (repressed) cultural image that it proposes to legislate against. What desires does that image satisfy? What fears does it raise?

In this chapter, I consider what ideological work is being performed by the image of ectogenesis, as it appears—only to be erased—in the context of our contemporary debate about embryo experimentation. That question—posed by the paradox embedded in the Warnock Committee report—can be approached by way of a history of the image of ectogenesis.

A HISTORY OF ECTOGENESIS

Ectogenesis, as an idea, is not a contemporary innovation. In fact, the Warnock Committee's specific connection between futurology (invoked/denied) and the image of ectogenesis (invoked/denied) was itself anticipated more than half a century earlier. In 1923, the publication of a wildly successful futurological pamphlet triggered a debate over the social, political, and medical implications of ectogenesis. A wide variety of writers, including prominent feminist novelists, social reformers, doctors, philosophers, and sexologists, took up the issue first addressed in J.B.S. Haldane's *Daedalus, or Science and the Future*: the invention and sweeping social adoption of the ultimate reproductive technology, extrauterine gestation or ectogenesis.[9]

Originally an undergraduate essay titled "The Future of Science," then an address to the New College Essay Society and the Heretics Society at Cambridge, Haldane's futurological pamphlet was finally published in 1923 as the inaugural volume of a newly launched venture, the To-Day and To-Morrow series. The publisher, Kegan Paul, Trench, Trubner and Company, ancestor of present-day Routledge, also claimed a position at the cutting edge of intellectual and social developments. They launched the highly visible To-Day and To-Morrow series to provide "a stimulating survey of the most modern thought" on "many departments of life," among them "Marriage and Morals," "Science and Medicine," "Industry and the Machine," and "War," by "some of the most distinguished English thinkers, scientists, philosophers, doctors, critics, and artists."[10]

With the publication in 1923 of *Daedalus, or Science and the Future*, the series challenged the social reformers of that most reform-minded of eras, the 1920s, to define just what they thought the future should be. In the following six years, five more essays were published specifically responding to Haldane's essay: both to its central image of ectogenesis and to the principles on which it was based—the separation of sexuality from reproduction; the benefits for society and the individual flowing from scientific control of human nature; and the concept that our biological and social behaviors were not natural, but naturalized.[11] Indeed, these essays debated the meaning of the pregnant woman herself.

When any image achieves cultural centrality, Mary Poovey has taught us, something larger is being fought out: thus the nineteenth-century debate over the medical use of chloroform constituted "a discourse of sexual politics in which the female body [was] politicized

—differently, and in the service of divergent interests."[12] A similar dynamic characterized the controversy spawned by Haldane's *Daedalus*: the notion of technologized reproduction served as the occasion for a debate over the pregnant female body, which was politicized to serve a great variety of ideological interests. The divergent representations of ectogenesis in Anthony Ludovici's *Lysistrata, or Woman's Future and Future Woman* (1924), Norman Haire's *Hymen, or the Future of Marriage* (1927), Vera Brittain's *Halcyon, or the Future of Monogamy* (1929), J. D. Bernal's *The World, the Flesh, and the Devil: An Enquiry into the Future of the Three Enemies of the Rational Soul* (1929), and Eden Paul's *Chronos, or the Future of the Family* (1930) reflect the authors' different social allegiances, whether to patriarchy or feminism, to Nietzschean aristocracy or to Marxism, to heterosexuality or what we might now call the lesbian continuum.[13]

A moment of cultural convergence between the discourses of eugenics, maternalism (or pronatalism), and sex reform in the early 1920s provides the historical context for this debate within which the technology of ectogenesis was constructed to serve various ideological agendas. Since its inception in the early years of the new century the eugenics movement had advocated a balance of positive incentives to reproduction by the fit and disincentives to reproduction by the physically or mentally unfit, but in the 1920s eugenicists began to stress policies of negative eugenics in response to a perceived threat of degeneration. Advocating direct intervention in reproduction to control the population and improve the species, eugenicists built on three important advances in scientific knowledge following the discrediting of the Lamarckian theory of acquired characteristics in the 1880s: (1) the notion that "environmental reforms could only have a limited effect [and] only selective breeding could improve quality"; (2) the new interest in a mathematical (or biometric) approach to problems of heredity; and (3) the rediscovery of Mendelian genetics, which led to a genetic approach to problems of population size and quality.[14]

Maternalism, or a pronatalist approach to population planning, extended this eugenic emphasis into the home, resulting in a reconceptualization of the role of the mother and of motherhood, in response to contemporary concerns about the (intellectual and physical) health and increase of the population. By the 1920s, reproduction came to be thought of as woman's duty to the "race"; as Weeks observes, "Child rearing was no longer seen as just an individual moral duty; it was a national duty" (*Sex, Politics, and Society*, 127). For working-class mothers, this meant a surge in new social organizations dedicated to improving the lot of mothers and babies; for the middle-class woman, it often meant the duty not only to reproduce oneself, but to help as unofficial monitors of others' mothering.

A new interest in achieving a scientific understanding of the whole range of sexualities and sexual behaviors led to the movement for sexual reform that gained importance throughout the 1920s. The field of sexology, founded in the early years of the century by Iwan Bloch and Magnus Hirschfeld, gained credibility in England with a research program viewing sexuality as a natural, rather than a constructed, physical activity that could be understood through biologically based experimentation.[15] As sexology attracted followers, some sexual taboos were (briefly) relaxed; female sexual pleasure, birth control, and even homosexuality became acceptable topics of conversation. But other constraints increased, Leonore Tiefer points out, most prominent among them the dominant analytic role of scientific discourse, and particularly biological and medical discourse, in setting research priorities, as well as in arriving at explanations of phenomena.

Pronatalists and sexologists converged in 1928 with the founding of the World League for Sexual Reform, whose self-proclaimed task was to promulgate "a new legal and social attitude (based on the knowledge which has been acquired from scientific research in sexual biology, psychology and sociology) towards the sexual life of men and women."[16] The League's platform reflected a seemingly paradoxical combination of maternalism, sexual emancipation, and sexual-purity agendas.[17] Yet the seeming paradox reflected the consequences of a commitment to sexual emancipation for women and men: because of the implications of sexual behavior for women, this had to include the empowerment of mothers. And because of the importance of sexology, such empowerment more often than not was advocated in scientistic terms.

The sexologists' consolidation would culminate in the Third International Congress of the World League for Sexual Reform, in 1929 (a year before the publication of Eden Paul's *Chronos*, the final pamphlet in the ectogenesis debate). This Congress convened a diverse group of radical and free-thinking leftist intellectuals concerned less with feminism than with sexual and social enlightenment. Yet as the proceedings, published in 1930, reveal, they arrived at a shared program affirming female sexual emancipation only by obscuring deep divisions within their constituency over eugenic measures ranging from contraception to sterilization to infanticide.[18] Enacting a scientifically endorsed split between female sexuality and maternity, the Congress affirmed (even emancipated) the former by ceding the latter to the domain of scientific and medical professionals.[19] The same conflicting social trends that split the League marked the decade of the ectogenesis debate: a new stress on the importance of motherhood, agitation for female sexual emancipation, and an interest in direct intervention in population control advocated by eugenicists.

The converging discourses of the 1920s produced "wildly utopian and scientistic" interventions into individual and collective sexual and reproductive behavior, Jeffrey Weeks has observed: such indeed was the nature of the ectogenesis debate (*Sex, Politics, and Society*, 124). Catalyzed by Haldane and engaging one author after another throughout the 1920s, the debate illustrates how the domaining of ideas occurs in and shapes our representations of reproductive technology.[20] By tracing the different positions motivating the contributions to the debate, we can identify cross-currents and conflicts within modernity, a discourse that has for too long been taken as seamless. Confirming that there is more to the modern than "the uniform logic of rationalization, repression, and masculine domination," such a project will also show that the boundaries of modernism and postmodernism are more porous than they are often taken to be.[21] Finally, an understanding of the history of ectogenesis can uncover the desires and fears that its image paradoxically both represses and releases in the present day.

HALDANE'S *DAEDALUS*

The focus of *Daedalus, or Science and the Future* is the potential of modern biology to reshape society through what Haldane calls biological inventions: "The establishment of a new relationship between man and other animals or plants, or between different human beings, provided that such relationship is one which comes primarily under the domain of biology rather than physics, psychology, or ethics" (42). Ectogenesis, with its dramatic potential to reshape human relations, is one such biological invention. Listing the six most important biological inventions of all time, Haldane deduces from them a general observation that will apply to ectogenesis as well: "The biological invention . . . tends to begin as a perversion and end as a ritual supported by unquestioned beliefs and prejudices" (49).[22]

Haldane chooses for the centerpiece of *Daedalus* an essay "on the influence of biology on history during the 20th century" which tells the story of the development of ectogenesis—or extrauterine gestation (56–57). Ironically, the narrator, a "rather stupid" Cambridge undergraduate of the twenty-first century, is modeled on Haldane himself, who was an undergraduate when he first drafted the essay that became *Daedalus*. That the undergraduate should be able to summarize the development of this complex biomedical technology suggests how thoroughly the innovation has been naturalized. That he should show no shame or self-consciousness in his narrative attests to the widening gap between sexuality and reproduction produced by the scientific challenge to the "natural" gendered body. Finally, that the narrative of its development makes it seem, to the reader, inappropriate to describe this unusual *reproductive* technique as a "perversion" (a term normally reserved for deviant sexual behavior) not only attests to the modern separation of sexuality from reproduction but also confirms Haldane's argument that biological innovations are inevitably accepted so thoroughly that they seem part of the normal and proper order of things. Indeed, this scientific breakthrough is reconstructed retrospectively from a future in which the discovery has been so completely accepted that it seems unremarkable: "As we know, ectogenesis is now universal, and in this country less than 30 per cent. of children are now born of women" (65).

The prophetic narrative is worth citing in full, not only because it offers an uncanny anticipation of the developments in reproductive technology since 1923, but also because it embodies in certain specific ways a familiar kind of popular science narrative:

> It was in 1951 that Dupont and Schwarz produced the first ectogenetic child. As early as 1901 Heape had transferred embryo rabbits from one female to another, in 1925 Haldane had grown embryonic rats in serum for ten days, but had failed to carry the process to its conclusion, and it was not till 1940 that Clark succeeded with the pig, using Kehlmann's solution as medium. Dupont and Schwarz obtained a fresh ovary from a woman who was the victim of an aeroplane accident, and kept it living in their medium for five years. They obtained several eggs from it and fertilized them successfully, but the problem of the nutrition and support of the embryo was more difficult, and was only solved in the fourth year. Now that the technique is fully developed, we can take an ovary from a woman, and keep it growing in a suitable fluid for as long as twenty years, producing a fresh ovum each month, of which 90 per cent. can be fertilized, and the embryos grown successfully for nine months, and then brought out into the air. (63–64)[23]

Haldane's fictional narrative is grounded in a fact. In 1880, Cambridge physiologist Walter Heape transferred "rabbit embryos from one animal to another on the point of a needle, in order to demonstrate both that implantation and gestation in another animal was possible and that there were no lasting effects on a superior female animal of mating with an inferior male."[24] Although the other episodes in this developmental narrative are fictional extrapolations from the state of scientific knowledge in 1923, they anticipate real developments in reproductive technology leading up to and including the development of in vitro fertilization. Dr. Alexis Carrel succeeded in keeping mammalian ovaries alive in tissue culture outside the body (1935); scientists experimented on the perfusion of fetuses and the culturing of placentas outside the woman's body (in 1958, 1962, 1968, 1969); a dispute over the fate of the frozen embryos of airplane-crash victims Mario and Elsa Rios received extensive media coverage (1984); extensive discussions of egg and embryo farming took place when the Human Fertilisation and Embryology Bill was debated in Parliament (1990); and the successful gestation and delivery of twins by a fifty-nine-year-old woman sparked a heated controversy over the transfer of embryos to postmenopausal women (in late 1993).[25]

Haldane's story of the development of in vitro gestation parallels the actual story of the development of in vitro fertilization, as told in Dr. Robert Edwards's autobiographical account. Both narratives move from successes in animal embryology to advances in human embryology: Edwards moving from mouse, rabbit, and monkey eggs and embryos to human eggs and embryos; Haldane predicting a course from embryonic rats and pigs grown in culture, to a human ovary grown in culture medium and used as source of eggs; a human embryo grown in culture; a human ovary cultured as egg producer indefinitely; and finally, successful human embryo culture, for a full gestational term.

Haldane's fictional narrative of ectogenesis echoes the classic progressivist picture of laboratory life, for its portrait of scientific discovery is constructed as a cumulative, collaborative process of challenges met and surmounted.[26] Suppressing both error and conflict, the tale has a narrow focus: the experiments on animal and human subjects through which a set of procedures was refined that ultimately make possible in vitro gestation, or ectogenesis. Contemporary cultural studies of science describe such a narrative as a scientific origins myth:

> Under the guise of telling people "where we come from," origins stories tell people "who we are." . . . Internalist histories of science . . . have this character: they claim that the discoveries of modern science reflect the pinnacle of human progress, and that the progress science represents is lodged entirely within scientific method. They tell

us "who we are": people who use scientific rationality to achieve progress in social life—including, of course, in inquiry.[27]

Not only does Haldane's narrative possess the coherence and tele-
ological certainty of an idealized portrait of scientific progress, it is also
explicitly framed as myth.

Beginning with its titular allusion to the legendary genius in "experimental genetics" who produced the first human-animal hybrid, the
Minotaur, built moving statues, crafted wings that permitted human
beings to fly, and as a punishment was imprisoned in the labyrinth he
constructed for King Minos, *Daedalus* relies on mythology to structure
its central point: that biology has the pivotal importance in the twentieth century that in the nineteenth century was held by physics and
chemistry (47–48, 10). Haldane personifies this institutional victory as
the defeat of Prometheus ("the chemical or physical inventor") by
Daedalus ("the first to demonstrate that the scientific worker is not
concerned with gods") (44, 48). With the victory of Daedalus over Prometheus comes a change in the nature and implications of scientific
knowledge: "Every physical and chemical invention is a blasphemy, every biological invention is a perversion" (44). Haldane's precise language (the shift from "blasphemy" to "perversion") emphasizes that
the site of transgression has changed: whereas physical and chemical
discoveries flout God's laws, biological discoveries disrupt normative
categories for sexual behavior, in particular the normalized gendered
body.

How does ectogenesis challenge what has hitherto been conceptualized as "natural sexual behavior," and what are the implications of
that disruption according to Haldane's futurological narrative? Not
only does ectogenesis interrupt, resituate, and scientifically usurp the
ordinary gestational consequences of human sexual intercourse, but
by enacting that "separation of sexual love and reproduction which
was begun in the 19th century and completed in the 20th," ectogenesis
changes the character of the "old family life [that] had certainly a good
deal to commend it" (65). Now, artificial interventions are required to
carry out relations that were previously "best in the instinctive cycle,"
most notably suckling, which now depends upon injections of placen-
tin (66). The undergraduate acknowledges that the changes in human
relations resulting from ectogenesis are not *all* to the good; its "effect
on human psychology and social life [is] . . . by no means wholly satis-
factory" (65). However, the overall tone of the undergraduate's potted

narrative of scientific progress is laudatory. It praises ectogenesis for enabling major eugenic advances in human breeding, particularly in the face of the increasing threat of species degeneration, and it approvingly details the process of selection required of those wanting to become ectogenetic mothers (66–69).

As Haldane formulates it, ectogenesis is linked to both positive and negative eugenics.[28] Deployed in response to a fear of inevitable species degeneration, ectogenesis becomes the centerpiece of a program of population control and species regeneration. Sanctions against ectogenetic breeding by the socially disadvantaged combine with incentives toward ectogenetic breeding for the socially privileged, producing improvement in the human species within the unthinkably rapid timespan of one human generation. With tongue in cheek, Haldane even suggests that ectogenesis may enable us to circumvent the clumsy political route to social change: "In the future perhaps it may be possible by selective breeding to change character as quickly as institutions" (69). Yet joking aside, although Haldane applauds ectogenesis for promising a useful control over the human reproductive process, he finds its meaning not in the process itself, but in its social deployment: the extent to which it is used to serve, or to abuse, the interests of the collective population. This suspended judgment reflects his conviction, as a socialist scientist, that extrascientific forces ultimately shape the meaning of scientific practice, so that even an oppressive procedure can ultimately have emancipatory results: "The tendency of applied science is to magnify injustices until they become too intolerable to be borne, and the average man whom all the prophets and poets could not move turns at last and extinguishes the evil at its source" (85).[29]

ECTOGENESIS AND SEX WAR

Daedalus looks cheerfully ahead to a future in which the invention of ectogenesis enables the control of human reproduction, the improvement of the human species, and finally the emancipation of mankind, but Anthony Ludovici's *Lysistrata, or Woman's Future and Future Woman* (1924) provides a gloomy mirror image of that inaugural vision of ectogenesis. Ludovici launches a virulent attack on scientific and industrial modernity, graphically picturing a malevolent science using the "body-despising values" of modern culture in an elaborate program of feminist domination and species degradation.

A translator of and lecturer on Nietzsche, Ludovici calls for a transformation of values in *Lysistrata*, in an argument both circular and illogical. Feminism encourages women to put the demands of the body second to the desires of the mind; feminism is an offshoot of modern "body-despising" values; so, in order to cure the human species of the ills of modernity (primarily "body-despising values"), feminism must be abolished. Woman's body thus takes the brunt of the social and ethical forces of modernity. As the *Times Literary Supplement* reviewer observed, "[According to Ludovici,] the greater part of the suffering resulting from defective corporeal equipment falls to the lot of woman."[30] Arguing that humanity must "put woman back in her place—which is only another way of saving the world," Ludovici advocates a return for woman not just from the public to the private sphere, but to the childbed (107).

Why is a recontainment of woman's activity from the sphere of public production to the private realm of reproduction so important to Ludovici? As its title suggests, in *Lysistrata* sexual reproduction is the contested practice in a modern sex war. In Ludovici's view, the converging forces of modern urban industrial mass culture, modern science, and modern feminism threaten to reconstruct not only production but reproduction.[31] Ludovici agrees with Haldane that in each arena, artificiality is on the increase, but he disagrees with his assessment of the implications of that increase. Whereas Haldane imagines the course of scientific progress as leading to a "life more and more complex, artificial, and rich in possibilities to increase indefinitely man's powers for good and evil," Ludovici sees scientific modernity taking a course that is increasingly evil, because increasingly artificial (20). Scanning a city street, finding there people who wear glasses and false teeth, are born with the assistance of forceps, take laxatives, and feed their babies artificial infant formula, Ludovici rails at the "highly standardized" nature of modern life (11). This tendency to standardization and artificiality, he argues, is the product of the combined forces of modern industrial production, modern science, and modern feminism.

First step in Ludovici's paranoid narrative of scientifically produced feminist mastery is the modern cultural valorization of mind over body, signaled by the increasing proliferation of "artificial aids": spectacles, processed foods, obstetrical forceps, and chloroform, cow's milk formulas for infant feeding. The next salvo comes with "an increasing emancipation of girls and women from domestic arts and duties . . . accompanied by the intensive manufacture of every kind of condensed, preserved, compressed, and synthetic food"; then follows an increasing segregation of men in the military and "an increasing assertion of the rights of females in every branch of industry, commerce, and the professions" (84–85). Finally, there is a "revolt against cohabitation" (anticipating trends in contemporary reproductive medicine): "The whole act of fertilization will be consummated in the surgery, just as vaccination is now" (88). As Ludovici forecasts it, the resulting feminist society, deeply sexually oppressive, only avoids the "periodical slaughter" of male infants by discovering "a means of determining the sex of the ovum," and limiting male births to one-half of one percent, for breeding purposes only (95).

Just as they agree that artificiality is the center of this new, scientifically controlled world but disagree on the implications of that fact, Ludovici and Haldane both represent human reproductive science working in tandem with animal husbandry, but again interpret this parallel in dramatically different ways. Haldane uses the examples of animal husbandry to support his assertion that biological inventions begin by repelling us as perverse and end up being accepted as natural parts of human social behavior. Ludovici, in contrast, uses the animal connection to raise the specter of devolution. Haldane sees the processes of infant feeding and sexual reproduction as amenable to human cultural construction and reconstruction, but Ludovici excoriates the "artificial feeding of infants" with cow's milk as "essentially a modern invention" that testifies ominously to the increasing feminist dominance in modern life.[32] Ludovici illustrates this point with an image loaded with sexual and racial anxieties: "Place a human baby at the dug of a cow, a goat, or an ass, as you sometimes see them placed in semi-civilized countries, and what is it that you immediately feel? The sight is an offense to the eyes, a humiliation of our racial pride" (70).

Ludovici revises the image of ectogenesis slightly to emphasize a similar set of devolutionary anxieties: he figures the gestation of human fetuses not by machines but by animals to stress the threat of degeneration he sees in ectogenesis:

> At first . . . this will occur by again enlisting the cow or the ass into our service . . . in the early days of extra-corporeal gestation, the fertilized human ovum will be transferred to the uterus of a cow or an ass, and left to mature as a parasite on the animal's tissues, very much as the newborn baby is now made the parasite of the cow's udder. And, with this invention, we shall probably suffer increased besotment, and intensified bovinity or asininity, according to the nature of the quadruped chosen. (91–92)

Envisioning an ectogenesis that threatens to transgress the boundaries between races and species, Ludovici articulates the complex anxieties elicited in the 1920s by the emergence of modern mass culture, the new industrial methods of Taylorism and Fordism, and the feminist movement.

But it is not only Ludovici's terror at female domination that makes *Lysistrata* such an interesting response to *Daedalus*. In contrast to the linkage between modern scientific objectification and masculine mastery, such as would animate Julian Huxley's story "The Tissue-Culture King" only two years later, Ludovici constructs a vision of science that links it to modern feminism. Ludovici's epigrammatic reference to the "quack cure of feminism" reveals his strategy of using gender to reconfigure modern science as a debased, commodified servant of mass industrial production and consumption (79). In this connection, we might adapt Andreas Huyssen's observation about mass culture in *After the Great Divide*: "The problem is not the desire to differentiate between forms of pure science and depraved forms of applied science and its co-options. The problem is rather the persistent gendering as feminine of that which is devalued."[33]

Whereas Haldane's assessment of ectogenesis reflected his commitment to a socialist future, Ludovici's reflected his fear of a feminist one. In the years that followed, the interventions of Norman Haire (a sexologist), Vera Brittain (a feminist novelist), Eden Paul, M.D. (a socialist eugenicist), and J. D. Bernal (a Marxist molecular biologist and crystallographer) would continue this tradition of using ectogenesis as the ground on which to contest quite other issues. In each case, as I will sketch out in what follows, a reconceptualization of the relations between fetus and gestating woman would serve as the fulcrum of predicted future change.

ECTOGENESIS AND SEXOLOGY

Sexologist Norman Haire inserts an assessment of ectogenesis into his wide-ranging analysis of problems of "sex life," *Hymen, or the Future of Marriage*.[34] Unlike the members of the Warnock Committee, who refuse to predict beyond a certain "time horizon," Haire asserts his right to engage in prophecy. He bases this "right" not on "omniscience or infallibility," but on the possession of "a fair amount of intelligence, a certain capacity for objectivity which one does not meet in one's fellows as often as one could wish, a high ethical standard, a well-developed social sense, and a real desire for the increase of human happiness caused by the removal of unnecessary causes of suffering" (96). Haire's little book thus explicitly constitutes itself on the borderline between pure science and popular science, the scientific elite and the general public, and the present and the future, if not between modesty and overweening pride.

Although Haire's assertion of his intelligence, objectivity, ethical standards, social sense, and altruism may fail to convince contemporary readers, *Hymen, or the Future of Marriage* does contain some insights into future social practices. Not that Haire got everything right: he forecast early marriages, polygamy, enforced sterilization for those unfit to breed, legalization of infanticide and euthanasia, and ectogenesis. But Haire seems to have sensed the pulse of changes now in progress, predicting the universal practice of contraception, the legalization of abortion, and a changed attitude toward sexual "abnormality," consisting in equal parts of prevention and tolerance (80–93).

Haire's vision of ectogenesis in *Hymen* reflects his institutional positioning as a sexologist. It avoids the mechanistic focus of Haldane and Ludovici, echoing instead the eugenic and interspecies emphases of his sexological research, dramatizing how genetically valuable fetuses could be brought to term by being transferred between women, even between species. Haire also offers disconcerting new perspectives on two contemporary biomedical techniques—surrogacy and organ donation—by framing them as ectogenetic practices:

> It is probable that young embryos—of good heredity on both sides— will often be removed from the uterus of the original mother and grown in the uterus of other women who volunteer for the service, or perhaps even in the uterus of other animals. Such transferences have already been successfully carried out by the experimental biologist, and this method of *Ectogenesis* is likely to be applied sooner or later in the human being. Ectogenesis may go even further. In the case of the accidental death of a woman particularly suited for parenthood, her reproductive glands may be transplanted into a female animal which had already had its own reproductive glands removed. Artificial fertilization with human spermatozoa would then give rise to a human embryo which would be incubated and brought to birth by the animal host. (87–88)

Haire's curious qualification—"a woman particularly suited for parenthood"—reveals that as a typical sexologist who privileges the biological over the social or psychological, he is interested less in women's experience (whether of sexuality, reproduction, or mothering) than in women's genes. This confirms Tiefer's observation that "the historic domination of sex research by biologically oriented theorists (studying, for the most part, animal subjects) has blinded sexologists to the bias such domination has imposed on the field" ("Feminist Perspective," 22). We might even speculate that this woman is particularly suited for parenthood because she combines "good genes" with the ultimate passivity of a corpse.

Haire's optimism that the technical problems of ectogenesis will ultimately be solved embodies that mixture of chauvinistic pride and hubris characteristic of newly consolidated scientific fields:

> Only a few months ago a French scientist [Dr. Serge Voronoff, of Paris] announced that he had succeeded in performing such a transplantation and fertilization, in a female monkey; and, at the moment of writing, the birth of a human child from the simian mother is eagerly (if somewhat skeptically) awaited. But whether this particular scientist has succeeded or not, there can be little doubt that the thing will be achieved sooner or later. (*Hymen*, 88)

The measured assessment of analysts in the 1980s and 1990s, ranging from the Warnock Committee to feminist theorists, that ectogenesis is improbable (because it costs too much and is too technically demanding) contrasts with Haire's practiced, confident assertion that the march of scientific (eugenic/ biological) progress will continue. In Haire's view, ectogenesis is not just a technique for enabling a fetus to grow outside a woman's womb; it is a technique for enabling a new medical/scientific field to grow (and flourish) outside medicine and biology. The institutional self-interest of Haire's assessment of ectogenesis demonstrates that biomedical innovations are not judged in a vacuum but rather assessed in relation to the institutional agenda(s) they serve, as well as to their cultural and social context(s). Moreover, it prompts us to keep in mind the institutional and sociocultural context when we weigh the implications of new biomedical technologies.

Ludovici and Haire approached ectogenesis from the broadest social perspective, the former fearing it as a feminist plot and the latter welcoming its potential to emancipate women and improve the species; however, Vera Brittain and Eden Paul thought of ectogenesis more specifically in relation to the changes it might make in individual and family life. Only J. D. Bernal conceptualized ectogenesis without considering its impact on women, either socially (like Ludovici and Haire) or personally (like Brittain and Paul). Instead, Bernal focused on the fundamental structure of ectogenesis: partnership between a human being (in this case, a fetus) and a machine (in this case, an artificial uterus). Although the implications of Bernal's work are closest, in a sense, to our own time, Eden Paul's brief discussion of ectogenesis in *Chronos, or the Future of the Family* (1930) and Brittain's more complex analysis in *Halcyon, or the Future of Monogamy* (1929) raise important issues that were not addressed by J. D. Bernal's better-known vision of ectogenesis in *The World, the Flesh, and the Devil: An Enquiry into the Future of the Three Enemies of the Rational Soul* (1929).[35]

Like Haldane and Bernal, Eden Paul combined socialism with science: before World War I he was a fervent socialist, and after the war he joined the Communist Party.[36] As a medical doctor, he advocated the conjoined use of eugenic and social reforms to better the human lot, for "unless the socialist is also a eugenicist, the socialist state will speedily perish from racial degradation."[37] His allegiance firmly fixed on maintaining the quality of the human race through changing times rather than on enforcing a stable system of morality, Paul acknowledged philosophically what Ludovici bemoaned: the disappearance of traditional family life under the joint impact of feminism and sexual reform. In *Chronos* he diagnosed "the present trend of civilized social life" not as the "revival of matriarchy" feared by many of his colleagues, but rather as a movement toward "a balance between the sexes," of which "that wider movement 'feminism' is merely one expression" (40).

Responding to the consequences of that modern shift in gender relations, Paul advocates in *Chronos* that the death of the family must be accepted, in order for a viable substitute for it to be found. If the "biological classification" of the family is fast disappearing, that does not necessarily mean that the home is doomed as well. Paul defines "home" as a nonbiological collective unit for the nurture and education of children, arguing that even in this era of shrinking population, rising divorce rates, and plummeting marriage rates, homes are needed to care for, and educate, the generation to come. But those homes will no longer be biologically based. Rather he advocates a system of "scattered homes" which will "grow in number as the decay of the family proceeds," and adds that it "will more and more become the recognized thing to send children to them when the age of suckling is over" (54).

Paul's stress on the cultural function of the home for education and social cohesion rendered him indifferent to its biological substrate. To the criticism that his plan disregards the crucial "ties of blood," Paul responds with impatience: "Sometimes, in my more revolutionary moods, I am inclined to think that they [ties of blood] are 'useless, dangerous, and ought to be abolished.' Anyhow, they counted for little with primitive man, and perhaps will count for still less when (if ever) man becomes truly civilized" (55). Instead, Paul shares with J.B.S. Haldane before him a humorous awareness of how innovations are soon naturalized. In *Chronos*, he gets that point across by telling his readers of the old lady from *Punch* "who, when asked to make a journey by airplane replied: 'I don't like these new-fangled methods. Give me the *natural* way of travelling, by rail" (50).

The expectation of a continual process of naturalization does not extend beyond social innovations to biological ones, however. In this respect Paul parts company with Haldane. Although Paul is impatient with a biological grounding for human relations in "ties of blood," there are some biological models he preserves—most notably the link between fetus and gestating woman. Like many of his feminist and sexological contemporaries, Paul acknowledges the appeal of ectogenesis: "There is no sexual reformer but must wish that woman could be freed from the slavery of child-bearing, and that our offspring might come into the world out of a broken eggshell" (35). But he concludes that at the moment, gestation seems to be part of that "period during which family life is essential, both for mother and for child" (34). With no further exploration, Paul dismisses the notion of ectogenesis regretfully:

> About the intra-uterine stage there is at present no option. In the last act of *Back to Methuselah*, Bernard Shaw toys with other possibilities, envisaged also in J.B.S. Haldane's remarks on "ectogenesis" in *Daedalus, or Science and the Future*, and in J. D. Bernal's *The World, the Flesh and the Devil*. . . . But "inter faeces et urinam nascimur" seems likely to be true for a long while yet. If we are talking "practical politics" we shall not contemplate the freeing of woman from this part of Eve's curse. (34–35)

The remark Paul makes in passing—that gestation is essential "both for mother and child"—forms the fulcrum of the assessment of ectogenesis by feminist, pacifist novelist Vera Brittain, in *Halcyon, or the Future of Monogamy* (1929). Brittain seems to have borrowed her rhetorical strategy from Dora Russell, who in *Hypatia, or Women and Knowledge* (1925) entered the debates on chloroform, infant feeding, and contraceptive and sex education, using scientific discourse as both cover and warrant for a liberal feminist argument: "[A] better understanding of psychology and physiology based on the discoveries of modern science—is bringing to the whole of life, but especially to sex-love, maternity, the rearing and education of children, joy and rapture and promise surpassing anything known to the purely instinctive life of the past."[38] Russell welcomes scientific knowledge as an enlightening advantage with which to counter religious fundamentalism and superstition (51). Although she never specifically addressed ectogenesis, we can hypothesize that her position would have been favorable.

Brittain's text shares Russell's strategic use of authoritative discourse for the subversive purpose of feminist advocacy, but here the authority is wielded not by science but—as we shall see—by high literary modernism. Brittain departs from Russell's scientifically grounded optimism; her position on ectogenesis is anything but simply positive. Instead, Brittain scrutinizes ectogenesis, as she does all of the other topics she addresses in *Halcyon*, in terms of its impact not on the species, nor on society in general, but on the specific experience of women *in relation to* their children. To understand the radical reach of Brittain's assessment of ectogenesis, we must understand its dual, equally radical conceptual reframing of literary modernism and modern scientific rationality.

Brittain's contribution to the To-day and To-morrow series, *Halcyon, or the Future of Monogamy* (1929) turned literary tropes to the task of futurological and scientific speculation, using the goal of fulfilled monogamous marriage as a warrant for advocating feminist demands. T. S. Eliot used the tale of the Fisher King (borrowed from Jessie Weston) to anchor his narrative of sexual regeneration in *The Waste Land* (1922). Virginia Woolf inverted and naturalized Eliot's mythic image in the birth scene of *Orlando* (1928), portraying not the Fisher King but the kingfisher as an emblem of sexual and reproductive happiness. In subversive homage to both of these texts, Vera Brittain uses the legend of the kingfisher to anchor a feminist plea for a reconstructed sexual and reproductive life for women in the modern era. In a frontispiece that explains the meaning of her title, she explains how the kingfisher myth anchors her reconstruction of modern sexual, social and literary life: "Alcyone, or Halcyone, in ancient Greek legend, a daughter of Aeolus and Enarete or Aegiale. When her husband Ceyx was drowned, Alcyone cast herself into the sea. The gods pitied the devoted pair, and changed them into birds, the alcyons or halcyons (kingfishers); it was believed that during the breeding-time of these birds, the sea was always calm" (*Halcyon*, 4). Brittain's use of the kingfisher motif attests to her sympathy for Virginia Woolf's feminist embrace of sexual and social multiplicity, as well as her distance from T. S.

Eliot's patriarchal anxieties. Rather than dying in order that the land and its inhabitants may live again (as in Eliot's poem), or living as both sexes (and both genders) in one mind and body (as in Woolf's novel), Brittain's kingfishers anchor a mythic vision of social harmony, a method of breeding that produces calm, for individuals and society.[39]

By the formal structure of her essay, Brittain telegraphs her intention to contest J.B.S. Haldane's biologistic model for reforming family life, in particular his central innovation: ectogenesis. Just as Haldane focuses in *Daedalus* on an essay by an undergraduate of the future, "on the influence of biology upon history," Brittain's focus in *Halcyon* is an excerpt from a work of "social and moral history destined to be published in the far-distant future" (5–6). Yet the authors, and hence the authority, of the two histories of the future differ dramatically. Whereas Haldane's is written by a somewhat dim male Cambridge undergraduate, Brittain's is written by Dr. Minerva Huxterwin, a brilliant woman professor, holder of the "Chair of Moral History" at Oxford University, and the female embodiment of the combined wisdom of Huxley and Darwin (6).

Like J. D. Bernal's work *The World, the Flesh, and the Devil* (also published in 1929), Brittain's essay charts a course of human development leading to perfection. Yet whereas Bernal narrates the physical development of a "perfect man," Brittain's Professor Huxterwin writes the history of the moral development of a whole society. From the post-Victorian era (1900–1930) through the "period of Sexual Reform" (1930–1975) and the period of "Scientific Progress" (1950–2000) to the final "Triumph of Voluntary Monogamy" (2000–2030), Huxterwin's essay charts how a series of feminist legislative provisions bring about increasing biological, sexual, legal, and social autonomy for women.

Anticipating late-twentieth-century critiques of the repressive hypothesis, Brittain's essay adroitly, paradoxically, calls into question a central tenet of the sexual reform movement: the distinction between repressive monogamy and liberating sexuality. Brittain argues that sexual diversity, if enforced, can be repressive, whereas voluntary monogamy in the context of female autonomy can be liberating. As Stephen Heath would point out half a century later: "The much vaunted 'liberation' of sexuality . . . is thus not a liberation but a myth, an ideology, the definition of a new mode of conformity (that can be understood, moreover, in relation to the capitalist system, the production of a commodity 'sexuality')."[40]

Halcyon also negotiates a feminist response to science that can stand as a valuable model for feminist scholars today. Donna Haraway has defined as a central challenge faced by contemporary feminism the creation of a workable response to modern science: one that does not either engage in Luddite rejection or wholesale, uncritical embrace.[41] Brittain's essay exemplifies such a complex, nuanced response, for Huxterwin approvingly cites Bertrand Russell's observation, in his contribution to the To-day and To-morrow series: "As science advances, more and more things are brought under human control" (53). And her history chronicles approvingly the scientific advances that have increased marital happiness: the wireless, television, the cinematograph, the "so-called 'home talkies' apparatus," and the perfection of high-speed air travel, which enables greater mobility and thus increases the happiness of monogamous couples (54–61).

Despite her affirmation of these techno-scientific advances, however, Brittain is alone among the respondents to Haldane's *Daedalus* in imagining the possibility of resistance to scientific progress. To Haldane's query, "whether there is any hope of stopping the progress of scientific research," Brittain responds with a resounding yes (5). In her discussion of "the attempts which began about 1950 to separate the sexual love of men and women from their generative processes," Brittain presents a powerful image of resistance to the inexorable progress of science (75). In Minerva Huxterwin's feminist moral narrative, the scientific development of ectogenesis is enacted, assessed, found wanting, redesigned in response to social requirements, and ultimately almost abolished.

Her narrative of the development of ectogenesis initially parallels Haldane's, only to depart from it significantly:

> [In] 1971, . . . only twenty years later than the date predicted for this event by Haldane, an ectogenetic girl was successfully reared through the embryonic stages and brought to "birth" in Monet's laboratory. Heedless of the world-wide ecclesiastical *furore* let loose by this triumph, the leading ectogeneticians of England, France, Germany, Russia, and the United States at once went rapidly ahead with the rearing of the embryonic children supplied by a small but slowly increasing number of co-operative parents. (76)

As Huxterwin describes it, the initial social impact of ectogenesis led to the rejection of old, nonscientific mothering practices. Thus the "intelligent 'mothering' of small children by their own parents," advocated by the Infant Welfare movement, was dismissed as "an out-of-date relic of sentimental Georgianism" (77). But as the numbers of ecto-genetic children increased, the scientific experiment that their creation effectively embodied began to produce disturbing data. "Scientific motherhood" no longer seemed to produce the best offspring, and the problematic developmental course of ectogenetic children led to the reevaluation both of currently accepted childcare theories and of the technology that such theories had so eagerly embraced. "[The] first laboratory-grown children . . . suffered as much psychologically from lack of individual parental affection as they gained physiologically through being selected from the best stock. The majority of them, indeed, though most carefully exercised, dieted and exposed to sunlight, dwindled away and died about the fifth year" (77). To Haldane's stress on genetic makeup or heredity, Brittain counters with a powerful corrective stress on interpersonal relations, or environment. This commitment to the social as well as the biological produces a wider range of options in response to ectogenesis: the tragic conclusion to the ectogenetic experiment leads not to scientific attempts to engineer-in artificial parental affection, but rather to a reassessment of ectogenesis itself. Since there is no coercion (neither social, political, economic, or scientific/technological) to maintain the investment in ectogenetic technology, Huxterwin's parents of the twenty-first century choose rather to avail themselves of another recent invention: "Kettmann's system of muscular and digestive control [that] rendered childbirth painless and pregnancy definitely pleasurable" (77). They return to "natural methods of reproduction," retaining ectogenesis only in a highly modified version, changed to reflect the new understanding of the crucial role of mothering:

> In the case . . . of the few marriages where normal pregnancy was exceptionally inconvenient to the wife, or would involve a long separation from her husband, recourse was had to the expedient,
> subsequently familiar under the name of Dickensian Pre-Birth, which Sir Frederick Benedickens invented in 1984.
> By this operation, now comparatively common, the fertilized embryo is removed from the mother a few weeks after conception, without damage to itself or deterrent effect upon her ability to conceive afresh, and is grown for eight months in Kolinovski's Gestative Solution. The majority of such children, however, are returned to the mother immediately after their first complete exposure to the air for the purpose of artificially induced lactation, and in no case are they allowed to remain in the laboratory for more than a year after "birth." (77–78)

In this narrative of ectogenesis, Brittain anticipated such contemporary techniques as embryo transfer and induced lactation, used by Maggie and Linda Kirkman in their 1988 sister-sister surrogate motherhood.[42] Moreover, unlike Eden Paul, Brittain explored thoroughly the importance of gestation to both mother and child, even to the extent of dramatizing how resistance to ectogenesis could spring from widespread social acknowledgment of the importance of the gestational relationship for the health of the future child.

Brittain also understood the complicated nature of that relationship for the gestating woman. Anticipating our current moment, when gestating woman and fetus are increasingly constructed as deadly adversaries in a pitched battle, Brittain mapped out some truce terms. She imagined the social and scientific negotiations with which the needs of mother, father/husband, child, and society could be addressed, and even to some extent satisfied, by balancing social and scientific interventions, rather than by giving in wholly to the demands and constraints of science.

Brittain's essay also shares the failings of its era. Reflecting her liberal individualist politics, it is cast entirely in the language of free choice. As Professor Huxterwin tells it, the parents of the twenty-first century make their reproductive decisions without social or scientific pressure. Her model offers no analysis of how resistance might emerge in a context such as that imagined by J.B.S. Haldane, where social coercion is applied to enforce eugenic regulations. Moreover, Brittain's utopia assumes (and requires) that science has been repositioned as one of a plurality of discourses, rather than the dominant one. When social contraindications to ectogenesis emerge, they have sufficient authority in themselves to prompt a reevaluation of scientific practice. Such a reconceptualization of the relationship of science to society remains to be accomplished in our day, and was far from fact in Brittain's. Her narrative of resistance to scientific innovation appeared at a time, and in a social context, when the discourse of science reigned supreme. This scientific hegemony explains why when Naomi Haldane Mitchison reviewed *Halcyon* for the *Eugenics Review*, she did not praise its vision of thoughtful negotiations between scientific and social interventions, but rather its portrait of scientific progress. Asserting (in response no doubt to her own personal misgivings) that the "monogamic millennium" "only comes about through much struggle and thought," Mitchison applauded Brittain for describing "the political and scientific events of the future, none of them very remote from what is actually happening now . . . partly this new era is to be reached through various biological discoveries, all quite easy to believe in."[43] Mitchison casts her praise of Brittain's "little book" in the discourse of scientific rationality: "How cruel and darkened, still more how silly, we late Georgians seem! How easy it would be for us to be rational in thought and deed, and make a happier world."[44]

Despite these shortcomings, it is impressive indeed that Brittain could imagine in *Halcyon* a world so dramatically different from her own. Her contribution to the ectogenesis debate maps out for us today a way of thinking about gestation that both affirms the experiences of the parties involved (gestating woman, fetus/child to be, and father/husband) and acknowledges the needs of society at large. Moreover, she formulates a way of responding to science that respects its powers without ceding the authority of other ways of knowing. Brittain's vision of gestation as a *relationship* stands as a powerful counterpoint to the other visions in the ectogenesis debate.

J. D. Bernal's book *The World, the Flesh, and the Devil: The Future of Three Enemies of the Rational Soul* was published in the same year (1929) as Vera Brittain's *Halcyon*, and it shares the strategy of narrating the stages in a march toward human perfection. Yet whereas Brittain's pamphlet traces the stages of progress toward a feminist future, Bernal not only ignores feminism, but sidelines women as well. An X-ray crystallographer and molecular biologist, Bernal was a celebrated polymath whose futurological speculations and extramarital wanderings shocked the staid Cambridge environment where he began his career, according to Gary Werskey (*The Visible College*, 80–81). Yet Bernal's science was always linked to a concern for social justice.

To the ectogenesis debate, Bernal's *World, the Flesh, and the Devil* contributes a fable about the possibility unleashed by this new technology for machine-gestation: the creation of a "perfect man such as the doctors, the eugenists and the public health officers between them hope to make of humanity" (42). Bernal's premise is simple: human evolution will continue, and it will ultimately include the machine. "Sooner or later, the useless parts of the body must be given more modern functions or dispensed with altogether, and in their place we must incorporate in the effective body the mechanisms of the new functions" (42).

Ectogenesis is the first step in the three-stage process that will produce a "man" who is perfect because he is freed from the sensory, motor, and biological constraints of the human body. This scientific evolution will begin in an auto-experiment seemingly modeled on the work of J.B.S. Haldane himself: "Sooner or later, some eminent physiologist will have his neck broken in a super-civilized accident or find his body cells worn beyond capacity for repair. He will then be forced to decide whether to abandon his body or his life. After all it is brain that counts, and to have a brain suffused by fresh and correctly pre-scribed blood is to be alive—to think" (42–43). Bernal maps out the progress toward perfection that follows this revelation that the mind can exist independent of the body. First will come the "larval" stage, beginning "as Mr. J.B.S. Haldane so convincingly predicts, in an ecto-genetic factory," and dedicated to the concerns of the body, including love-making and "perhaps incidentally . . . reproductive activity" (45). The second, or "chrysalis," stage begins when the man leaves his body, "whose potentialities he should have sufficiently explored," and undergoes "a complicated and rather unpleasant process of trans-forming the already existing organs and grafting on all the new sen-sory and motor mechanisms" (45). This medical/technical procedure will be followed, Bernal predicts, by "a period of re-education in which he would grow to understand the functioning of his new sensory or-gans and practice the manipulations of his new motor mechanism" (45–46). The third, and final, stage will be reached when the perfect man emerges from his chrysalis "as a completely effective, mentally-directed mechanism" who is "physically plastic in a way quite tran-scending the capacities of untransformed humanity" and thus "able to extend indefinitely his possible sensations and actions by using suc-cessively different end-organs" (46). That this "perfect man" exem-plifies accepted modern medical and scientific practice, by taking the male as the model for the human, may be so obvious as to require no emphasis. But what are the implications of Bernal's fusion of human (read "male") and machine?

The first two stages in the creation of Bernal's "perfect man" de-pend on two characteristic modern developments: the functional spe-cialization built into modern industry, which through Fordism and Taylorism enforced "behavioral engineering that treated the body as a machine," and the act of objectification central to the modern scien-tific method, embodied in the "decontextualization of the object [of scientific knowledge] and its insertion into a laboratory situation."[45] Like the king's tissues and the axolotl, the brain of the perfect man is removed from its sheltering context (the body, in this case) and sub-jected to scientific procedures in the new environment of the labora-tory. These modern attributes also appear in the final stage of Bernal's perfect man, which recalls works of science fiction from the 1920s such as Julian Huxley's "Tissue-Culture King" and Flagg's "Machine

Man of Ardathia." Mechanized for "scientific rather than aesthetic purposes," the apotheosis of this process of joining man to machine resembles the machine man of Ardathia:

> Instead of the present body structure we should have the whole framework of some very rigid material. . . . In shape it might well be rather a short cylinder. Inside the cylinder, and supported very carefully to prevent shock, is the brain with its nerve connections, immersed in a liquid of the nature of cerebro-spinal fluid, kept circulating over it at a uniform temperature. The brain and nerve cells are kept supplied with fresh oxygenated blood and drained of deoxygenated blood through their arteries and veins which connect outside the cylinder to the artificial heart-lung digestive system—an elaborate, automatic contrivance. . . . The brain thus guaranteed continuous awareness, is connected in the anterior of the case with its immediate sense organs, the eye and the ear—which will probably retain this connection for a long time. The eyes will look into a kind of optical box which will enable them alternatively to look into periscopes projecting from the case, telescopes, microscopes and a whole range of televisual apparatus. The ear would have the corresponding microphone attachments and would still be the chief organ for wireless reception. (47–48)

The perfect man ends as a brain in a beaker, having begun as a fetus in a machine uterus. The basically electrical function of the brain having been discovered and its engineering mastered, "co-operative thinking" based on a sort of inter-brain networking becomes possible: "Connections between two or more minds would tend to become a more and more permanent condition until they functioned as dual or multiple organisms" (52).

How, according to J. D. Bernal, would the creation of this perfect man change society? Like Julian Huxley before him, Bernal predicts that the future state will be a pharmacracy, characterized by a division between "the scientists and whose who thought like them—a class of technicians and experts who would perhaps form ten per cent. or so of the world's population," and "the rest of humanity" (90). Bernal envisions that this scientific state, founded on "the colonization of the universe and the mechanization of the human body," will stay in power not through force majeure, but through the softer coercions of biopower: "Psychological and physiological discoveries will give the ruling powers the means of directing the masses in harmless occupations and of maintaining a perfect docility under the appearance of perfect freedom" (89). A new, scientifically created and enforced class system will emerge: the populace will be divided into "altered and the non-altered humanity" (89–90).

This vision of control stops short of eugenic "cleansing." Not for humanitarian reasons, however, but scientific ones: Bernal understood the importance of genetic diversity and the maintenance of a varied gene pool. The nonaltered human beings will still be important, he chillingly predicts, not in themselves, but for their genetic material. Thus even when space exploration becomes possible, and the new, mechanically enhanced and "better organized beings" engage in the colonization of other galaxies, the nonaltered humanity will be preserved. The world will be "transformed into a human zoo, a zoo so intelligently managed that its inhabitants are not aware that they are there merely for the purposes of observation and experiment" (95). Such a plan, Bernal ironically observes, "should please both sides: it should satisfy the scientists in their aspirations towards further knowledge and further experience, and the humanists in their looking for the good life on earth" (95).

A SYMPTOMATIC READING
OF THE ECTOGENESIS DEBATE

What makes the ectogenesis debate in the 1920s important for us today? These little-known pamphlets, published as futurological speculation in the To-day and To-morrow series, can illuminate the contemporary issues at play when we talk—or refuse to talk—about ectogenesis. To tease out the ideological work currently being done by the image of ectogenesis, we can perform on it what Jane Gallop has called a "symptomatic reading": a reading that "squeezes the text tight to force it to reveal its perversities," a reading that when applied "to culturally powerful texts can be a tool for diminishing their power," but when turned "on culturally marginalized texts . . . may . . . have at least the momentary effect of promoting the text to canonlike status."[46] A symptomatic reading of the image of ectogenesis will shift from reading it as a symptom of the individual who created it, J.B.S. Haldane; to seeing it as symptomatic of its era, a 1920s obsessed with reproduction and sex; to analyzing what the resurgence of the image of ectogenesis means in our own time. Of course, if we read ectogenesis as symptomatic of J.B.S. Haldane the individual, we must also understand the extent to which Haldane was himself symptomatic of his era and spoke "from within a field of combat."[47] But with that caveat, we can specify the unconscious issues that were being fought out (for him and for his era) on the terrain of reproductive science: gender identity; the conflict between identification and control; and the boundaries between human, animal, and machine. Although a full symptomatic reading, like the psychoanalytic process in which the technique originates, is necessarily interminable, we can sketch out the lines along which it might proceed.

Haldane's character was, by all accounts, powerfully paradoxical. He exhibited, from his earliest years, a remarkable capacity to split himself off, objectify himself, in the service of science. As his sister Naomi Mitchison recalled, he and his father, the distinguished physiologist John Scott Haldane, "worked together from the time when Jack was old enough to remember a formula or hold a test tube."[48] When he was only three years old, his father was already using him as an experimental subject, drawing samples of his blood. At thirteen, he donned a diver's suit to serve as guinea pig in one of his father's experiments with "the bends": the suit, which was too large, leaked, and young Jack Haldane was pulled to the surface in a suit full to the waist with water. Later, while still a teenager, he again served as the experimental subject in a study of the effects of fire damp on the lungs, descending at his father's side into a mine and reciting Shakespeare until he collapsed.[49]

By the time Haldane had become a don at Oxford, the habit of experimenting on himself was deeply ingrained, a habit he would later document in his essay, "On Being One's Own Rabbit."[50] Attempting to confirm experimentally his father's theory of the functioning of carbon dioxide in human blood, Haldane and his colleague Peter Davies applied "the Golden Rule (which he learned from his father): 'To test on ourselves first that which we would have others do.'"[51] Using themselves as guinea pigs, Haldane and his colleague followed the principle later elucidated by his biographer, Ronald Clark: "Neither a dog nor a rabbit, nor any experimental animal other than man can 'tell you if he has a headache, or an upset of his sensators of smell, both of which I obtained as symptoms during these experiments'" (*J.B.S.*, 59).

This practice of auto-experimentation has a long history in scientific practice, extending back to Aristotle and reaching forward to our own time, where it was central to the in vitro experiments of Dr. Robert Edwards.[52] But Haldane's auto-experimentation stands out, both for how it exhibited his own transgressive notion of science, and for the remarkable cross-identification that characterized it. What Haldane found, during his auto-experiments, was as likely to be a self-revelation as a theoretical principle, a literary insight as a scientific truth:

> If I find out how to produce a certain change in the composition of my blood I want to know what it feels like to appreciate it as a fact of life as well as a fact of chemistry. Thus I regard it as interesting that, after taking the largest quantity of calcium chloride on record, I dreamt that Edward Lear had written and illustrated a life of Christ. It was a strange book, but not essentially irreverent. Unfortunately, the only detail of it which remains clearly in my memory is Pontius Pilate's moustache.[53]

Even in his last months, dying of cancer, Haldane "behaved as though the experience of dying was an interesting experiment, to be conducted as such," according to his friend Frank Crew. He responded to the news of his mortal illness characteristically, by composing a poem, "Cancer's a Funny Thing," which expressed in its opening couplet his lifelong defiance of boundaries, whether between literature and science, or life and death: "I wish I had the voice of Homer / To sing of rectal carcinoma."[54]

Ronald Clark, Haldane's biographer, interprets his practice of auto-experimentation as characteristic of the extravagant, irrational side of his character: "However useful it was, and of that there can be no doubt, there was a trace of exhibitionism about the way he spoke of such work, a flamboyance epitomised by his comment that the only way to test a chemical's reaction was to take ten times the dose listed as fatal in the British Pharmacopoeia" (*J.B.S.*, 61–62). Yet considered in conjunction with his invention of extrauterine gestation, this persistent positioning of the self as (scientific/social) outsider suggests more than simple flamboyance: a genuine curiosity about the experience of the Other.

Read symptomatically, Haldane's invention of ectogenesis reflects his passionate interest in, and desire for, alien experience. Haldane's friends and his first wife recall him as persistently interested in otherness, fired by a transgressive curiosity that extended beyond the boundaries of sex, species, even of animate and inanimate objects. As Charlotte Haldane recalled in her BBC-broadcast, "My Husband the Professor," he boastfully, zealously imagined a nonhuman life, taking on the role of "auto-guinea pig," even envying slugs their sex lives, "since they are hermaphrodites, and he thought they must have more fun than humans."[55] Others shared Charlotte's image of Haldane as a chimeric, boundary-defying creature. Ronald Fraser caricatured his "unhumanness" in the figure of Mr. Codling in *The Flying Draper*, "It was like being friends with fish, or a bird, or a half-human God." Haldane jokingly allowed himself to be caricatured as a fish, in fact, in the frontispiece to *Animal Biology*. (Ill. 13.) As Martin Case recalled to Lord Ritchie-Calder on the BBC, "Haldane identified himself with the persecuted minority, even to the extent of inventing minorities."[56] Inventing ectogenesis, Haldane takes on the reproductive position of woman, an imaginative act that combines subject with object, self with other, identification with control.

If Haldane spoke "out of a field of combat," what was that contested discursive field into which the ectogenesis debate erupted in the 1920s? How did the debate reflect the cultural wishes and fears then under circulation? As Jeffrey Weeks explains, it was an era obsessed with reproductive control and power. So-called *artificial* birth control became a crucial issue in the years from 1910 to 1929 for a number of reasons: the increased publicity given condom use during the First World War, which eroded anticontraceptive feelings; continuing high maternal mortality rates during the years between the wars; increasing awareness of the hardships faced by many mothers; new concern over the postwar population decline; continuing concern over eugenic issues; and perhaps most important, the fact that "by the 1920s technical advances did seem to open up the possibility of artificial [birth] control on a large scale" (187–188).

As women were gaining control of their reproductive powers through contraception, legislative control too seemed not far behind. With limited female suffrage won in 1918, women agitated for increased family allowances, better child welfare, a whole spectrum of social improvements. This new forceful feminist agitation no doubt raised a sometimes frightening specter of female power. Anthony Ludovici's link between feminism and reproductive autonomy, while resulting in a diametrically opposed assessment of ectogenesis than that of the other essayists, nonetheless articulated the sense they all shared that the balance between men and women was shifting. Debating ectogenesis in an era of seismic shifts in gender relations, Haldane and Ludovici, Haire and Paul, Brittain and Bernal, all responded to the concept of a technique for extrauterine gestation by defining, even renegotiating, the terms of the wider social field. They imagined the renegotiation of gender relations produced by ectogenesis as the foundation for sweeping change in other relations as well: between

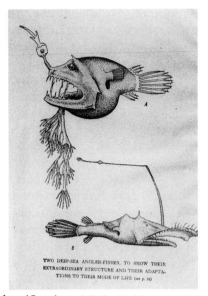

TWO DEEP-SEA ANGLER-FISHES, TO SHOW THEIR
EXTRAORDINARY STRUCTURE AND THEIR ADAPTA-
TIONS TO THEIR MODE OF LIFE *(see p. 12)*

13 Cross-species Identification: J.B.S. Haldane and Julian Huxley as Fish

the individual and the species, adults and children, human beings, animals, and machines.

Contesting these relations fiercely, the ectogenesis debate in the 1920s is also important, finally, simply because *it was a debate*. Unlike the Warnock Committee, which in 1984 declined to debate the pros and cons of ectogenesis, dismissing it rather as both unpractical and unpalatable, the contributors to the To-day and To-morrow series took ectogenesis very seriously. Rather than tabling it as a question until its scientific and technical challenges had been solved, they examined its implications and argued for or against it with a passionate respect for scientific innovation.

ECTOGENESIS AND THE CYBORG

What is the significance of ectogenesis in our own post-Holocaust, post-Hiroshima, postmodern era? What is its function, for example, when simultaneously invoked and denied in the Warnock Committee report? As I understand the image of ectogenesis, it functions as a strong defense against the anxiety raised by what Patricia Yaeger has called reproductive asymmetry: the fact that women and men contribute in unequal ways to reproduction. Drawing on Mary O'Brien's monumental work *The Politics of Reproduction*, Yaeger describes the different asymmetries that "make up the core of a modern reproductive unconscious: (1) the 'alienation' of men's sperm in the moment of coitus; (2) women's contributions to reproductive process as a form of transformative 'labor'; (3) the unequal division of this reproductive labor between the sexes; and (4) the fact that the child is, at least potentially, 'a value' produced by reproductive labor."[57] Yaeger argues that these asymmetries have produced the "obsessions with reproductive labor that pervade Western literature" (285). I argue that it is from this core of anxiety over human reproductive asymmetry that the image of ectogenesis speaks. As a fantasy, ectogenesis functions as an attempt to abolish the asymmetry, or at the very least to even it out. In ectogenesis, the notion is, gestation takes place in an artificial uterus; both the man's sperm and the woman's egg are equally alienated from their very moment of union, the moment of coitus having been replaced by in vitro fertilization. And with in vitro gestation, woman's gestational contribution to the reproductive process is replaced by man's scientific contribution, as the one who has built and who monitors the machine and its contents.[58] Finally, the unequal division of birthing work, historically favoring women, is inverted. Even in the era of Lamaze the act of giving birth has largely, if not wholly, been the woman's contribution. Yet with the advent of what Aldous Huxley dubbed "decanting," the act of giving birth falls under the dominance of science, historically gendered male.[59] The child is now "a value" produced not by reproductive, but by medical scientific labor—any putative ownership thus falling not to the mother, but to the engendering, laboring, delivering power of science. Expressed, repressed, and ultimately dismissed by the Warnock Committee as (implicitly) an irrelevant futuristic fantasy, still the notion of ectogenesis is profoundly, immediately relevant. Read symptomatically, it speaks of the deepest fears and wishes of our cultural unconscious: the fear of female procreative dominance, and the male wish to usurp and monopolize reproductive power.

Calling for a "poetics of birth," Yaeger has identified a conflict within feminism between essentialist celebrations of the birthing body, and postmodern suspicion of both "mothering and its metaphors" (293). The latter position is most dramatically embodied for her by Donna Haraway's image of the cyborg, "not of woman born" and "suspicious of the reproductive matrix" (293). Attempting to forge a position that will allow feminists to claim our reproductive experiences without being imprisoned by them, Yaeger proposes an audacious compromise: "Why not reimagine pregnant women as cyborgian ciphers rather than tossing out gestation and parturition—as if these biological events were responsible for the dreams of Cartesian man?"[60]

Although I share Yaeger's frustration at this impasse in our feminist response to reproduction, the implications of the ectogenesis debate lead me to find this compromise position unworkable. The cyborg can offer no escape from the "dreams of Cartesian man," I argue, for *the cyborg is not a cipher*. It has a history, extending back to the early years of the twentieth century. And it has an origin, revealed by the ectogenesis debate in the early years of this century. If we return one last time to J. D. Bernal's book *The World, the Flesh, and the Devil*, we can reconstruct the postmodern cyborg's modern lineage clearly and discover why it won't work as an emancipatory model for a feminist response to birth.

Beginning with an ectogenetic gestation and birth, a human being is created who receives prosthetic body extensions in the form of mechanical body parts, parts that are infinitely interchangeable and replaceable, enabling limitless sensory experience and intricate, sensitive, and flexible motor capabilities, but which require complicated training to use. This paraphrase of J. D. Bernal's contribution to the ectogenesis debate reveals that the Cartesian fantasy of freeing man (as mind) from the mortal body begins with the notion of freeing the fetus (as proto-man) from the female body. In short, *the cyborg originates in ectogenesis*. Foundational to the fantasy of the cyborg, with its denial of the mind/body link, is an earlier denial of the relationship between fetus and gestating woman.

The image of the cyborg has achieved such broad cultural prominence as to become "nearly canonical," as John Christie has observed. Yet recovery of the repressed relations between the cyborg and the earlier notion of ectogenesis can prompt us, as feminist theorists and cultural critics, to reevaluate the cyborg as an emancipatory model for women in our time. Donna Haraway used the term "cyborg" in her influential essay "A Manifesto for Cyborgs" (1985) to refer to the "cybernetic organism, a hybrid of machine and organism, a creature of social reality as well as a creature of fiction," that "changes what counts as women's experience in the late twentieth century."[61] Casting the cyborg as an escapee from the oppressive binaries of oedipal sexuality, positioned without fear at the intersection of different genders, races, and species, Haraway celebrated its role in unmasking "ideologies of sexual reproduction [that] . . . call on notions of sex and sex role as organic aspects in natural objects like organisms and families."[62]

Yet although Haraway acknowledged the possibility of a more oppressive cyborg, "the awful apocalyptic *telos* of the 'West's' escalating dominations of abstract individuation, an ultimate self untied at last from all dependency, a man in space," she never linked that image of ultimate independence to the anxiety actuating it: fear of the "reproductive matrix" itself, embodied by the fetus's ultimate dependence on the body of the gestating woman.[63] The ambivalence of the cyborg has thus seemed a product of postmodernism's notoriously shifting politics, spawned by the postmodern discomfort with positionality and identity.[64]

Yet "cyborg postmodernism . . . can also in certain aesthetic and political registers recapitulate and rhetorically rehearse, as well as move beyond, that emergent modernist moment which encountered woman, science, technology, and politics in the same frame."[65] I understand the ectogenesis debate as the originary and repressed stage in the modern confrontations and negotiations between woman, science, technology, and politics reimaged in cyborg postmodernism. Revealing a profound ambivalence about the gestating, birthing, *experiencing* female body and mind characteristic of the modern period, the ectogenesis debate managed and contained that ambivalence through the fantasy of the scientific birthing of a "perfect man": the cyborg.

Despite its emancipatory overlay, then, the fantasy/image of the cyborg is a problematic choice as a resolution of the feminist quandary about birth. We need to consider what it means to think of the pregnant woman as a cyborg, when the cyborg itself has links—via the shared origin in the fantasy of ectogenesis—to a number of troublingly oppressive contemporary representations of pregnancy: from anti-abortion images of the fetus/patient at risk in the hostile, dangerous environment of the pregnant woman, to advertising images that solicit consumer identification with a fantasy-fetus as passenger in a machine-as-gestating woman.[66]

THE PATENT PLACENTAL CHAMBER

In the summer of 1993, the *New York Times* reported that "a Philadelphia obstetrician has patented an artificial uterus that would suspend a fetus in a liquid environment until its lungs matured."[67] The patent for a "placental chamber" was actually filed by Dr. William Cooper in February 1991, only three years after the Warnock Commission report issued its recommendation paradoxically declaring ectogenesis to be "beyond the time horizon" of the inquiry, and thus outside the bounds of its predictive competence, while simultaneously recommending that it be outlawed. As the *Times* described the ectogenetic machine:

> Dr. Cooper's invention is a small chamber divided into upper and lower sections. The lower half would be filled with synthetic amniotic fluid, imitating the natural liquid that surrounds a fetus in the womb. . . . The fetus would float in the liquid solution of the artificial uterus's lower half. Its umbilical cord would reach through to the upper part, where the placenta would be spread over the horizontal shelf dividing the two areas.[68]

The illustration that accompanies the patent bears an uncanny resemblance to some of the early ectogenetic images, and familiarity with the early-twentieth-century debate over ectogenesis can illuminate this late-twentieth-century patent application. (Ill. 14.) Like its precursors, Dr. Cooper's image of ectogenesis functions as a strong defense against the anxiety raised by reproductive asymmetry. A passage in the patent application suggests some of this anxiety:

> It is one object of this invention . . . to provide a system which provides a fetus with an artificial environment which mimics the baby's prebirth environment.
> At a gestation period of about 10 weeks, the baby and placenta are substantially independent of external hormonal input, and the only maternal functions are (1) oxygenation of the placenta and (2) nutrition of the infant with removal of subsequent waste products.[69]

The depersonalizing and disarticulating processes integral to this representation of ectogenesis express a desire to deny the subjectivity of the pregnant woman. With the mother-fetus relationship reduced to the functions of oxygenation and nutrition, the artificial environment of the placental chamber seems ready to take over for the already dehumanized "prebirth environment."

The patent also arguably serves two additional, overlapping, but by no means identical, discursive positions: the anti-abortion ("right-to-life") position, and the demands and self-constructions of a newly professionalizing medical field. The inventor, Dr. William Cooper, head of the Christian Fertility Institute in Easton, Pennsylvania, seems to be advocating an alternative to abortion with his "placental chamber,"

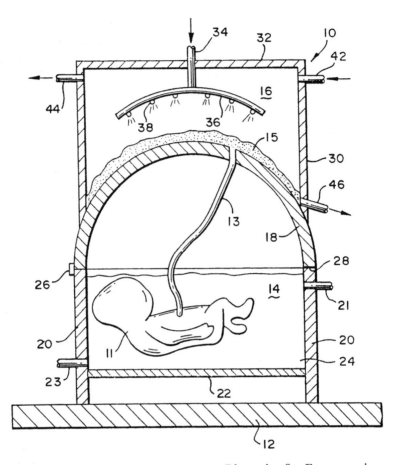

14 Placental Chamber: A Contemporary Blueprint for Ectogenesis

which he describes as capable of "supporting the life of a prematurely-born baby."[70]

However, the text of the patent can also be read as an assertion of the expertise of a newly consolidating medical field, "fetal therapy." This new subspecialty of high-risk, intrauterine fetal medicine began in 1982, "when a small group of obstetricians and geneticists . . . decided that medicine was far enough advanced for them to start treating fetuses as patients."[71] The field shares with IVF a developmental trajectory moving from animal experimentation to human experimental medicine, and it promises to raise equally charged ethical and social issues. A recent example was the December 1993 case of

Tabita and Mercea Bricci. Doctors at St. Joseph Hospital in Illinois determined that the thirty-six-week-old fetus carried by Mrs. Bricci "was not getting enough oxygen or nutrients from the placenta," and they urged an immediate Caesarian section to "spare the fetus from almost certain death or mental retardation." When the Briccis refused to permit the Caesarian section, they were sued by the Cook County (Illinois) State's Attorney and Patrick T. Murphy, a court-appointed lawyer for the fetus. After their case was upheld by a panel of the Illinois Appellate Court, Mrs. Bricci delivered a seemingly healthy, if somewhat undersized baby boy. Ethicists grappled with, and seem to have been baffled by, the issues this court case raised: "Dr. Mary Mahowald, assistant director at the Center for Clinical Medical Ethics at the University of Chicago and a professor of obstetrics and gynecology, said the case demonstrates that although medical science is rushing into the future it is still fallible. . . . 'A good ethical principle,' Dr. Mahowald said, 'is when in doubt, trust the patient.'" Yet that is precisely what the new fetal therapy calls into question: Who is the patient? In their shift of focus from gestating woman-as-patient to intrauterine fetus-as-patient, these two discourses arguably converge, sharing the strategy and perhaps even the desire to abolish the agency of the pregnant woman and replace it with the agency of medical science.

In both its early- and late-twentieth-century incarnations, the ectogenesis debate illuminates the wishes and fears that motivate the scientific project of gaining control over reproduction and thus doing away with woman's asymmetrical reproductive role. In particular, it speaks to the desire to deny the subjectivity of the pregnant woman, and to end fetal dependency on her, generating instead the image of the cyborg. But the ectogenesis debate also suggests an answer to the natal impasse in which feminism finds itself now. For the varied views of ectogenesis held by the widely diverse group of writers engaged in the debate—socialists, conservatives, feminists, masculinists, liberals and Marxists, members of the religious right, and high-tech medical practitioners—teaches us that no technology or medical/scientific practice is debated in a vacuum. Both the technologies and those who assess them are haloed by their discursive and cultural contexts, and we must look to those contexts for the fullest understanding of the social implications of any technology, whether it is the hypothetical ectogenesis, or the actual contemporary technologies of embryo experimentation and fetal therapy.

three

Sex Selection, Intersexuality, and the Double Bind of Female Modernism

The aim is to control the reproductive process: the number of children that women should have, the timing of their conception, their sex. Almost everything seems to be controlled by some agency other than women themselves. This raises the basic question of who should have the control: scientists, religious establishments, the state, or women themselves? —R. P. RAVINDRA

In 1990, Baroness Mary Warnock caused a furor when it was reported that she felt a technique of prenatal sex determination could legitimately be used to maintain class distinctions: "Baroness Warnock believes hereditary peers should be allowed selective in-vitro fertilisation for male heirs to continue family lines. . . . The baroness said a statutory body would probably be set up to rule out frivolous applications for sex-selected babies. 'But I don't think an hereditary peerage is a frivolous thing,' she said."[1]

Baroness Warnock's view of prenatal sex selection emphasized its potential to perpetuate class and sex privilege. But others represented the practice in very different terms. In the same month as Baroness Warnock's notorious statement, the *Independent* reported the successful development of a technique of prenatal sex determination through embryo biopsy by emphasizing its benefit not to the male but to the female sex. The article specified that this procedure would increase the survival rate of female embryos: "Three women are now pregnant with female foetuses whom they know will be free of the 200 genetic defects that affect only males." Pointing out that "all this will be illegal" if Parliament bans embryo research, the article quoted one of the doctors who had perfected the technique: "The whole procedure is so rapid that we can biopsy a group of embryos in the morning, amplify the DNA and identify those that are male, and transfer selected female embryos to the mother in the evening."[2] The *Times* medical correspondent Aileen Ballantyne described how this "revolutionary treatment" made it possible to selectively implant desired female embryos in the womb: "The sex-check is done by removing a single cell from the embryo, which is created in a laboratory dish by fertilising the egg with the male partner's sperm. If female, the embryo is then transferred to the woman's womb, where it grows normally."[3]

Although such techniques for sex determination are still relatively new, the debate over their gender and class implications has a discursive history extending back over half a century.[4] One illuminating site of such debate is the fiction and popular science writing about sex determination and intersexuality of Charlotte Burghes Haldane, the first wife of J.B.S. Haldane. More than sixty years before the "sex-check" breakthrough, this feminist journalist and popular science writer published a novel exploring the social implications of prenatal sex selection. As she explained her idea for the novel: "It occurred to me to wonder what would be the effect on society if the human race could determine in advance the sex of its children. But to deal with this theme in fiction would mean taking a large imaginative stride into the future, since present biological research had not yet made a society built on such a postulate a practical possibility."[5] Published six years before Aldous Huxley's *Brave New World*, Haldane's novel was aptly titled *Man's World* (1926).[6] The novel portrays a society organized around prenatal sex selection, in which women are reduced to biology, categorized by their reproductive and sexual roles, and ruled by a coterie of racist white male scientists through a network of cybernetic surveillance and biological controls.

We can turn to the works of Charlotte Haldane to trace some of the history of the debate over prenatal sex determination and to discover what that debate reveals both about the meaning of these new reproductive technologies *and* about our cultural representation of them. In particular, Haldane's writings about prenatal sex determination and intersexuality bear the marks of her own conflicted feelings—as a modern woman writer—about sexuality and gender. Because she wrote when prenatal sex determination was still only a hypothetical technique, her works reveal the cultural and ideological work performed by the image of reproductive technology, namely, to manage conflicts over sexuality and gender in divergent ways, in conformity to different agendas.

As my opening juxtaposition of contemporary reproductive technology and early-twentieth-century literature suggests, in this chapter I have two seemingly divergent commitments. First, I want to tell part of the history behind our current debate about prenatal sex selection. I believe that it will be helpful to see how feminists in former days addressed the issues posed by those technologies. Second, I want to use Haldane's novel about sex selection as a test case for a different way of situating women writers in relation to literary modernism. I believe that it will be helpful in this discussion to see how contemporary feminist literary critics are reevaluating modernism and how contemporary feminist critics of science are reevaluating modern science.

A powerful, if often culturally unacknowledged, relationship exists between the creation of fiction and the construction of fact. Charlotte Haldane's representations of prenatal sex determination reveal this relationship, as the revisionary movements of female modernism and feminist critique of science converge analytically in their identification of one specific, and highly contested discursive site: the pregnant woman's body.

Because that site is the central focus of Haldane's novel, *Man's World* thus provides an appropriate point of intervention to reevaluate the modern woman writer's relation to modern science. Charlotte Haldane's popular science writings and her autobiography, like her novel, all illuminate our cultural representation of the three discourses circulating in this chapter: modernism, modern science, and feminism. It will be useful to look first at the broader cultural context for those writings.

SEXOLOGY: RETHINKING SEX

Biology was a powerful language in which to express the changing sense of the human condition at the turn of the twentieth century, as we have already seen in the writings of Julian Huxley and J.B.S. Haldane. In the wake of Darwin and Mendel, biologists increasingly called on theories of evolution, degeneration, and heredity to draw parallels between the development of the human race and the development of other species. Not only were a whole range of fictions, from children's stories to novels and short stories, suffused with what we would now think of as a naturalist's consciousness, but the dominant discourse of biology spawned other, more focused discourses linking the reproductive histories of individuals to the fate of the human race: eugenics, sex reform and sexology, and vocational mothering.

The growth of the eugenics movement was fueled by the anxieties about sex difference widespread in Britain during the interwar era, anxieties arising from the wartime deaths of so many young men in World War I and the new cultural position of women following the war. These anxieties led to an interest in policing the boundaries of sex difference and controlling the outcome of reproduction. In particular, the population decline caused by battlefield deaths renewed popular interest in aggressive policies of negative and positive eugenics: incentives for those who suffered from hereditary diseases to refrain from having children, as well as economic and social encouragement given to reproduction by those who were deemed socially and biologically superior.[7]

Consolidation of the new scientific field of sexology, whose origins lay in the eugenics movement, was celebrated in 1929 at the Third International Congress of the World League for Sexual Reform. In his presidential address, Dr. Magnus Hirschfeld lauded Francis Galton, founder of the Eugenics Society, as one of the "great pioneers" who began the study of human sexuality as part of a broad scientific project: "the discovery, description, and analysis of the 'laws of nature.'"[8] Hirschfeld claimed for the sexologist the fields of "sexual biology, sexual pathology, sexual ethnology ['the sexual life of the human race from prehistoric times up to our own'] and sexual sociology."[9]

The new discipline of sexology produced new taxonomies of sexual pathology, scientific responses to the anxiety about gender boundaries following the First World War. Prominent among them the category of the "intersex," a term referring to a condition of morphological indeterminacy or gender nonconformity ranging from the physiological (i.e., individuals who appear outwardly to correspond with one sex, but whose sexual organs are appropriate to the other, or who possess organs of both sexes) to the social (i.e., "mannish" behavior in a woman; "womanish" behavior in a man).[10] Some biologists in the 1920s used such categories to argue for universal potential bisexuality, but when the biological data were applied to society by the sexologists, they were typically used to police gender boundaries and to shore up the differences between so-called normal male and female behavior and appearances.[11] The cultural freight of these supposedly objective scientific categories becomes apparent when we consider that social concern over intersexuality was greatest in the years immediately following World War I, whereas by 1931 scientific opinion began to hold that intersexuality neither implied deviance in sexual behavior nor threatened society. According to H. G. Wells, Julian Huxley, and G. P. Wells, writing in *The Science of Life*, "The importance of these unfortunate eccentrics is enormously overrated, both by their adversaries and themselves, and more than half the evil of their misfortune is due to such exaggeration. . . . The true intersex is physically mal-formed, a thwarted being in that respect but not a perverted one."[12]

Common to these discourses inspired by late-Victorian biology was the new sense that human sexual behavior, and perhaps even the hu-man species itself, might be capable of human (re)construction. Yet despite this shared assumption, neo-Darwinists, eugenicists, sexolo-gists, and sex reformers ignored the constructed nature of the gender distinction that shadowed the sex distinctions with which they were concerned. Sexologists and sex reformers alike built their progressive social program on a set of uninterrogated gender-role assumptions, expressed in scientific discourse: the notion of innate, biologically based, immutable differences between the sexes in terms of sexual be-havior; the notion that ideal sexual relations are not only heterosex-ual, but male-dominant/female-submissive; and a notion of the normal female role comprised wholly of motherhood.[13]

The notion that the sexed human body was subject to scientific (re)construction had a complex relation to the valorization of the ma-ternal role. Whereas contraceptive education was an important goal of both the sexologists and the sex reformers, who supported the scien-tific separation of sexuality from reproduction, many sexologists also vigorously promoted the notion of vocational or "racial motherhood," holding that to produce healthy children was a woman's duty to the nation and the race.[14] This biologically based construction of woman's role exalted motherhood and maternal agency but at the price of fe-male diversity. The result was a balance between feminist and masculi-nist eugenic rhetoric and an apportioning out of agency between women (who experienced it in motherhood) and men (who wielded it in science).

As befitting that parallel process, a late-Victorian and early modern fascination with biology reflected the vicarious reproductive fantasies that have played a prominent role in utopian and dystopian fictions since Thomas More's *Utopia* (1516) and Francis Bacon's *New Atlantis* (1626), both of which experimented with notions of selective breed-ing.[15] H. G. Wells drew on that utopian tradition to explore different modes of reproduction in his turn-of-the-century writings, but the no-tion of scientifically controlled reproduction really achieved cultural prominence, as we have seen, with Haldane's *Daedalus, or Science and the Future* (1923). Indeed as *Daedalus* and Julian Huxley's "Tissue-Culture King" both attest, the climate of scientific enthusiasm in the first three decades of the twentieth century made the notion of reproductive control seem, if not yet fully practical, still something more than just a fantasy. Whereas novelists and social theorists were considering the social impact of scientific techniques for controlling reproduction, biologists and embryologists were experimenting with techniques of mechanically induced parthenogenesis, embryonic grafting, and tissue and cell culture.[16] When the fascination with biology in the early 1920s led to investigations of the mechanism of sex determination, the time was ripe for an upsurge of scientific and social interest in prenatal sex selection.

PRENATAL SEX SELECTION

Charlotte Haldane expressed this broad sense of biological possibilities in a range of writings from science journalism and social analysis to the novel that is the focus of this chapter, and finally to her autobiography, *Truth Will Out*. Six months after J.B.S. Haldane predicted extrauterine gestation in *Daedalus*, she published a column on the social impact of prenatal sex selection in the *Daily Express*.[17] "The Sex of Your Child" exemplifies the double bind of a woman who desires to improve the lot of women, yet who is confined to the maternal sphere and the legitimating discourse of modern science if she wishes to address issues of female agency.[18] (Ill. 15.)

In "The Sex of Your Child," Haldane proposes to explore current developments in the scientific project "to probe, reveal, and ultimately to control the forces behind the phenomena of existence." Inspired by Julian Huxley's writings on sex determination, the column illustrates the uneasy alliance of feminist and scientific discourses in the modern period. It forecasts the issues to be raised by a technique for prenatal sex determination, asking whether "a new discord [will] creep into domesticity when the sex of a child as yet unborn must be determined by consent of its parents." But because it focuses not on the private but the public implications of such a revolutionary change in human reproductive patterns, Haldane's column forecloses any serious discussion of the question it explicitly raises.

Despite her passing concern for how it would ignite gender-based conflict even within the private home, Haldane's response to the possibility of prenatal sex determination is basically positive. The development of such a technique, she predicts, would result in major changes in the social organization of race and gender, changes she envisions in a peculiar mix of masculinist and feminist scenarios. The ability to produce male babies would give crucial support to colonialist projects, but

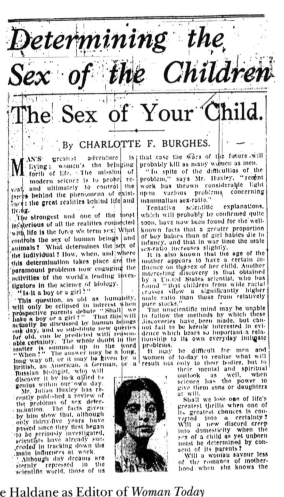

15 Charlotte Haldane as Editor of *Woman Today*

it would also eradicate gender-based social inequality. Once the sex of children could be chosen before birth, Haldane predicts, the "surplus women" problem would disappear. Fewer women would be born, and thus it would—over time—be possible for every woman to become a mother.[19]

Haldane's focus on empowering mothers and broadening women's access to the experience of maternity produces these meliorist predictions, which ignore the links between gender and war, the implicit racism in the construction of the colonizing project, and the phallocentrism of the assumption that only women who are legally married may become mothers, whereas unmarried women are problematic because unpartnered with a man, and thus "surplus." Welcoming the new technology because it provides a biologically based control that can be exercised on both the individual and the group, the physiology and the psyche, Haldane argues that with prenatal sex determination it will be possible to organize society rationally, on a local and global level. Global relations between empire and colonies will be orderly, and there will be a workable balance of marriageable men and women. Not only will this technique solve social problems, according to Haldane, but it will somehow solve biological problems as well. Haldane predicts a future free of the sexual boundary confusion that produced so much cultural anxiety in the interwar years: with the advent of prenatal sex determination, there will no longer be the problem of the "intersex," children will no longer be burdened with parental resentment for being "the wrong sex," and society will no longer be forced to deal with unacceptable sexual anomalies.

INTERSEXUALITY AND VOCATIONAL MOTHERHOOD

The "problem" of the intersex preoccupied Charlotte Haldane not just while writing "The Sex of Your Child," where she imagined its biomedical solution through the development of a technique of prenatal sex determination, but in *Man's World*, where she considered how resistance to techniques of prenatal sex determination could actually produce intersexuality, and in her pronatalist tract, *Motherhood and Its Enemies* (1927). Because intersexuality was one of the prominent worries addressed by the advocates of prenatal sex selection in the 1920s and 1930s, we need to understand what Haldane meant by the term if we want to evaluate her representation of sex selection.

Central to Haldane's argument in *Motherhood* is an attack on the "intersex" woman, whom she blames for [male-female] "sex-antagonism."[20] Haldane defines the "intersexual woman" as one who "[deviates] more or less markedly from the feminine form towards the anatomical and psychological characteristics of the masculine sex" (158). Despite this largely biologistic construction, in *Motherhood and Its Enemies* she focuses most on the behavior of intersexual women, which she blames for reducing the social standing of mothers: "In the past few years, particularly since the war, when [intersexuals'] advertised activities threw into the background the less spectacular exertions of mothers . . . their influence has grown alarmingly" (156).

Constructing intersexuality as the social and rhetorical opposite to vocational motherhood, *Motherhood and Its Enemies* reconstructs female sexuality as almost totally limited to reproduction. Thus Haldane condemns the intersexual woman not for her homosexuality, but for her failure to have children, which—reflecting the anxieties of her era—she sees as posing a grave threat to the survival of the human race. Although *Motherhood and Its Enemies* has been described as an "antifeminist classic," in fact it advances certain limited feminist positions, among them access to contraception (for married women), subsidized motherhood, and increased research into, and use of, anesthetics in childbirth. Yet the arguments for those feminist interventions rely on a disturbingly antifeminist distinction between mothers and other women, whom she defines as "abnormal": spinsters, war-workers, suffragists, and feminists.[21] The distinction between women is necessary to Haldane's argument, because she grounds tendentious social observations in the authorizing discourse of biology. Making maternal status rather than sexuality the ground of her argument, she blames competition among mothers and childless ("intersexual") women, rather than sexual oppression, for the particular "problem of sex" that concerns her, "the [debased] position of motherhood in the modern world" (146, 8).

According to Jane Lewis, such a privileging of motherhood may have been virtually unavoidable at the time: "[Few] women would have dared to speak against motherhood when the quality and quantity of population was considered to be of such great national importance."[22] Yet although Haldane idealizes motherhood in this pamphlet and advocates a scientifically administered maternal bureaucracy, she does not put wholesale trust in science. In *Motherhood and Its Enemies* she also criticizes eugenics as a possible origin of racist and classist abuses: "As certain ordained and even lay preachers of eugenics prove, this science holds potentialities of great danger" (238). Haldane, bitingly, continues: "Let the class-conscious or race-proud individual . . . with a mere smattering of scientific knowledge, attain any influence in this matter, and those whom he fears or hates (the same thing) will fare hardly. One would require a certificate of psychological purity even in the case of certain scientists before one would entrust them with so dangerous a profession as that of human geneticist" (238).

Central to *Motherhood and Its Enemies* is an argumentative strategy Haldane shared with other women of her day: recourse to scientific discourse to advance her feminist goals. Despite her acknowledgment that science could enable race- and class-based abuses, she attempts to formulate a scientific basis for elevating the social position of mothers. The contradictions of this position were probably unavoidable for any modern woman writer. As Jane Lewis has observed, "By the late nineteenth century it was already necessary to demonstrate a scientific approach in order to gain full recognition. . . . The use of biological analogy, in particular, proved very popular in explaining all kinds of social problems."[23] The same difficult balancing act between the feminist desire for female agency and the discursive hegemony of the largely male medical profession set the boundary conditions for a debate over prenatal sex determination that followed the publication of Charlotte Haldane's essay "The Sex of Your Child."

THE SEX-DETERMINATION DEBATE

Just as J.B.S. Haldane catalyzed a debate over ectogenesis with his *Daedalus, or Science and the Future*, Charlotte Haldane's essay "The Sex of Your Child" led to a debate over the method and meaning of sex determination. The interchange began with a series of articles appearing in 1925 in the *Daily Express* and the *Sunday Express*, the two newspapers for which she wrote regularly. By 1925, the debate had spilled over from the popular press to the *Lancet*, illustrating the high level of interest in both lay and medical communities in the question of sex determination. Taking place during the years in which Charlotte Haldane was planning and researching *Man's World*, this debate suggests some of the novel's cultural sources and anticipates the novel's strategy of using science as the ground on which to contest broader social issues.

The debate began on 2 July 1924 with the report that a method for prenatal sex determination was nearly a reality. The *Daily Express* reported that a maternal blood test had been discovered that would disclose the sex of the foetus.[24] Eight days later, *Daily Express* followed up on that topic with a front page story: "Secret of Sex Control on the Verge of a Discovery." The column reported the prediction of the director of the Animal Breeding Research Department, interviewed at a Scottish cattle breeding conference in Edinburgh, that within the decade "the scientist will have obtained such information concerning the processes of sex determination and sex differentiation that it is by no means impossible that the means of controlling the sex of offspring will have been developed."[25]

By 13 July 1924, the question of the social impact of such a discovery began to emerge in a series of articles that anticipated by over half a century the media tumult following Baroness Warnock's fantasy of prenatal sex determination as a prop to the peerage. The *Sunday Express* ran a front-page story by the *Sunday Express* "Special Correspondent" titled "Baby Boy or Girl? M.P.'s Wife Able to Read Nature's Secret."[26] Mrs. Monteith Erskine, wife of a member of Parliament, announced her ability to determine the sex of her children before birth. The column reported that her husband, J. M. Erskine, bragged uxoriously, "Many an heir to title and great estates owes his very existence to the instructions bestowed upon his mother by my wife."

Mrs. Erskine advanced a model for sex determination that seems distinctly pro-feminist.[27] In her model the ovaries determine the sex, with the ovary on one side producing eggs that result in girl babies and the ovary on the other, eggs leading to boy babies: "The sex of every egg exists before it is fertilised. And the ovary on each side produces only eggs of one sex. The ovaries work alternately, and between their months of work there is a fruitless month. Conception is only probable when the right day of the right month is chosen for intercourse. And the sex of the infant conceived will depend upon which ovary has been at work last."[28] Mrs. Erskine's theory extrapolates from the role of the egg in sex determination to the role of women in society, in order to affirm the social superiority of the female sex. Moreover, women's superiority extends to their knowledge about reproduction as well. As she explains it, her technique of sex determination confirms the greater value of women's direct testimony about their reproductive experiences than the secondhand reproductive information of a male physician:

> Only a woman can really solve these sex questions. There are innumerable things which must be felt in person to be known or understood. Doctors and men of science have floundered hopelessly. . . . My knowledge of dairy farming and the breeding of cattle helped me in my discovery, but hardly as much as the talks I have had with women of every kind and of every position about their confinements and everything to do with their experiences of conception and of birth. These are talks that no man could ever have had.

Mrs. Erskine's "Startling Claim to Predetermine Sex" continues to merit column space in the *Daily Express* on 14 July, the following day, though only on page 9. But there is a difference in the tone of the article by this point: it hedges on Mrs. Erskine's assertion that women excel in their understanding of the birth experience. Although it observes that "women especially have been greatly interested in her claims," it asserts even more prominently that "one of the greatest Scottish gynaecologists" has endorsed Mrs. Erskine's claim that "man . . . possesses no sex potentiality . . . [that it] does not in the least depend on [the father] whether his child will be a boy or a girl."

Not surprisingly, the medical community soon took issue with this bald assertion of male powerlessness in the arena of sex determination. On 15 July 1924, "Baby Boys at Will?," an article in the *Daily Express* whose author is styled merely "an eminent gynaecologist," refutes Mrs. Erskine's theory as "one of the most ancient ones," which has had to be "discarded by embryologists and gynaecologists." Instead, the unnamed gynecologist describes as the two leading explanations for sex determination "war and nourishment," that is, the notions that (1) "the more powerful element tends to produce a child of the opposite sex," and that (2) "the sex is determined during the time before birth—perhaps by the general health of the mother." The article concludes, "It seems likely that [sex] determination lies . . . not in the female or male element alone, but in the reciprocal influence they exert, one on the other."

While the British popular press continued to explore this question during the summer of 1924, it was not until the following spring that the authoritative medical journals articulated their position on sex determination. Not surprisingly, the effect of their intervention was to reassert the intellectual and social dominance of medical science, in highly [male] gendered terms. *Lancet* entered the debate on 25 April 1925 in a brief article titled "Determination of Sex." The writer acknowledges the existence within popular culture of an appealing utopian vision of female reproductive agency and control:

> [Our] daily journals have of late been regaling their readers with semi-bowdlerized versions of recently expressed views on the subject, from which it would appear that women may not only "have children at their desire," but . . . have them, to order, of any sex that happens to suit their particular requirements at the moment. Such, at least, seems to be the physiological Utopia whose coming is dazzling the anticipatory imaginations of the general public. (877)

In face of this climate championing the determining role of the egg (and consequently the woman), the anonymous writer goes on to assert the opposite construction. The article reports that contemporary scientific findings "[put] the chances of sex determination to the account of the paternal contribution," and consequently to the "sex difference in the chromosome content of the gonocyte nucleus" (878). A final article in the *Lancet* on 27 June 1925, attempted to settle the question with the rather hedged opinion that although "the chromosome factor and the environment influence do not necessarily stand as opposed hypotheses," the chromosome factor is most likely to be the overriding one (1355).

What is at stake in this debate over sex determination? The work of biologist Anne Fausto-Sterling can begin to clarify the broader issues. Fausto-Sterling's summary of the history of sex determination recounts that early-twentieth-century biology decontextualized the human ovum from the maternal body and developed two conflicting models of embryological development: a model privileging the role of the egg cytoplasm in determining the nature of the embryo and a model privileging the determining role of the cell nucleus because of the presence within it of genetically coded material. There were powerful, if only implicit, gendered implications to this conflict within embryology: "One can also see in this struggle a debate over the relative importance of male and female parents in the determination of the offspring's characteristics."[29] The champions of the egg cytoplasm argued not only for the shaping role of environment but also, by extension, for the "special role of the female parent," a position for which they were at times stigmatized.[30] In contrast, the supporters of the dominant role of the cell nucleus argued for the determining role of the genetic material.

The controversy described by Fausto-Sterling disputes the locus of developmental determinism within the individual fertilized egg cell rather than at the (earlier) moment of conception and exemplifies the process of arguing from the microbiological to the social that also figures in the debate over sex determination in the *Daily Express*, the *Sunday Express*, and the *Lancet*. The medical privileging of the contribution of the sperm over the egg shades into a privileging of the genetic material in the nucleus over the cytoplasmic environment, and thus into a victory of paternal over maternal contribution to the sex of the developing embryo. The fantasy of female, nonmedical control over sex determination has been effectively routed.

The exchange of views between lay people and physicians concerned with sex determination ends in a consolidation of scientific and medical authority. That consolidation is affirmed when we realize that our current understanding of sex determination corresponds to the understanding put forth, tentatively, in the final article in the *Lancet*: although chromosomal and environmental factors are both important in sex determination, the chromosome factor is most likely to dominate.

Medical truth is not the only issue at stake here, of course. We have seen how Mrs. Erskine's model for sex determination is grounded on women's direct reproductive testimony. We can trace a line from that phenomenological feminist model of birthing lore directly to the feminist critique of reproductive medicine advanced by contemporary scholars Barbara Katz Rothman, Rayna Rapp, and Emily Martin.[31] With Mrs. Erskine's model, as with those examined by Rothman, Rapp, and Martin, at stake is not so much the correctness of the medical information as how these models anchor an entire way of perceiving the world and understanding one's position within it. When women abandon their constructions of the reproductive experience and accept the "truth" of the dominant medical models for conception, gestation, and parturition, they also forfeit this woman-centered epistemological and ontological gestalt.

The 1924–1925 debate over sex determination reveals that twentieth-century women were particularly positioned—even trapped or torn—between conflicting discourses and epistemologies. This double bind, recently understood as central to woman's experience of both modernism and modern science, is embodied in Charlotte Haldane's writings. A review of the feminist revision of modernism and the feminist critique of modern science can set the context for a reading of Haldane's autobiography and fiction.

RETHINKING MODERNISM AND SCIENCE

The term 'modernism' has generated a number of critical misunderstandings that feminists are now rectifying as we "[rechart] the modernist territory."[32] Critics have had a long-standing, problematic tendency to confuse modernism with modernity and modernity with masculinity. Yet if we disarticulate those terms, as Rita Felski has observed, "A once established feminist view of the modern as exemplifying a uniform logic of rationalization, repression, and masculine domination no longer appears quite so compelling."[33] Just as the modern appears more complex and less hegemonically masculinist than we once understood it to be, so too the masculinity-modernism conjunction appears less seamless, when we realize that modernism has for too long been "unconsciously gendered masculine" by literary critics.[34] The label "modernist" slides to scientificity by way of its elision of all that is female; art, aspiring to science, aims at detachment, objectivity, unsentimentality, and thus (implicitly) masculinity.[35] With the modernism/modernity/masculinity nexus called into question, we are now able to rethink the hitherto unexamined relations between modernism and modern science, which has itself been understood as an equally, and unproblematically, masculine territory.

Andreas Huyssen was one of the first to suggest that the modernist commitment to scientificity may originate in a masculine reaction-formation against feminized mass culture. His summary of the canonical modernist aesthetic—as autonomous, self-referential, individualistic, and ironic—reveals it to have an "experimental nature [that] makes it analogous to science, and like science it produces and carries knowledge."[36] Those same modernist aspirations to scientificity motivate a rejection of subjectivity and authorial voice and an embrace of the ideals of autonomy and aesthetic distance.

Deferring for a moment the problems with the construction of science as monolithic and masculinist, we can see that the modernist aesthetic of scientificity (understood as objectivity, abstraction, and detachment) was propelled—as Suzanne Clark has recently argued—by a recoil away from sentimentalism.[37] Until quite recently, that recoil was also seen as implicitly political. Scholars connected "modernist disruptions of realist narrative" with disruptions of the "structures of authority that support imperialism, bourgeois class hegemony, and the male dominated family."[38] Yet claims for the implicitly emancipatory class politics of the modernist aesthetic are undermined by its more problematic gender politics. As Clark observes, "Modernism practiced a politics of style, but it denied that style had a politics" (*Sentimental Modernism*, 5). At the base of the hierarchized oppositions foundational to modernism is a gender politics: art/society, author/reader, form/content, man/woman, science/religion, reason/emotion, objectivity/subjectivity, image/narrative, mind/body. Those oppositions are all organized by the devaluation of that which is feminized: society, the reader, content, religion, emotion, subjectivity, narrative, the body.

Given the implicit gender politics of canonical modernism, when the writer is a woman the question of the politics of style is far more complicated than previous critical assessments of modernism have suggested. It is "too simple," Rita Felski has observed, to assume that "the political value of a text can be read off from its aesthetic value as defined by a modernist paradigm, and that a text which employs experimental techniques is therefore more radical in its effects than one which relies on established structures and conventional language."[39]

The uninterrogated assumption that modernist aesthetics was ideologically as well as stylistically uniform has produced a misunderstanding of modernism. But new feminist mappings of modernism have rectified this error, revealing that modernism includes stylistic regions previously thought out of bounds, such as the sentimental, the autobiographical, the popular, the romantic. Thus Suzanne Clark finds in modernism "a doubleness as well as a double bind" for women: "Modernism is both caught in and stabilized by a system of gendered binaries: male/female, serious/sentimental, critical/popular. Upsetting the system—as women do—introduces an instability and reveals the contradictions" (*Sentimental Modernism*, 8).

The deconstructive turn that has complicated and enriched our understanding of modernism is now being extended to our understanding of modern science as well. Just as modernism is far more complex than was permitted by the narrow, scientistic, and aesthetically uniform construction of canonical high modernism, so too both the history and the current state of modern science are more complex than we have been able to see or willing to acknowledge.

According to Londa Schiebinger, since the scientific revolution of the seventeenth century, a central strategy in the institutional consolidation of scientific fields has been the exclusion of women and women's issues. If we reverse the field of her insight, we can envision a fuller history for all the sciences: one that includes the peripheral contributions of women cooks, botanists, astronomers, instrument makers, herbalists, mathematicians, entomologists, and the list continues. And if the history of science is more complex than we had hitherto thought, the current institutions and practices of science are also more diverse and internally contested. As John Christie and Sally Shuttleworth point out, "The principle Enlightenment image of science as empirical, positivist, politically reformist, ideologically emancipatory" currently functions less as an accurate model than "as an obscuring, historical abstraction."[40] The cultural studies of science, extending from the early work of Thomas Kuhn to more recent work by Latour and Woolgar, Keller, Rouse, Schiebinger, Haraway, and others, have drawn our attention to the differences *within* science, both between scientific fields and from one scientific era to another.[41] These scholars have illuminated the processes by which scientific fields as diverse as cell biology, primatology, and physics have constructed both the questions they ask and the artifacts they accept as facts in relation to the cultural and historical milieu. Their findings suggest that we should speak not of monolithic science, but of scienc*es*: a collection of dynamic "forms, practices and representations" that change in response to different local sites as well as over time.[42]

Taken as a plurality of competing discourses, these newly understood sciences will occupy a variety of genres as well. For, as Schiebinger has shown, the institutionalization of science involved a narrowing not only of its gender composition, but of its gendered generic categories as well:

> By the late eighteenth century, scientists and philosophers were championing a science stripped of all metaphysics, poetry, and rhetorical ornament. . . . Literature, which Claude Bernard called the "older sister of science," was to be distinct from science. It was banished from science under the disgraceful title of the "feminine." The equation of the poetic and the feminine ratified the exclusion of women from science, but also set limits to the kind of language (male) scientists could use. (*The Mind Has No Sex?* 158–159)

Because science, like modernism, has forged itself through the devaluation of the feminine, a crucial act of institutional self-creation has been the cutting of a divide between scientific practice and the feminized discursive fields of literature, and poetry in particular. If we reverse the field, we recuperate the rhetorically feminine for scientific discourse. The reliance on elements of the romance genre in Charlotte Haldane's scientific writing finds its origins in this modernist act of recuperative field reversal, as we shall see later.

As this survey of the fields suggests, contemporary feminist critics of science have joined with feminist literary critics reassessing modernism to move us beyond the critical misapprehension, inappropriate totalization, and gender bias that once shaped our understanding of modernism and modern science. If we now understand that science need not be monolithically masculinist in its discursive, epistemological, or ontological properties, we also understand that science, like modernism, has historically shaped itself thus in reaction against the anxieties aroused by the discursively, epistemologically, and ontologically feminine. Feminist work on literary modernism and feminist critique of science converge to reveal that both fields were constituted, in the cultural imaginary, by the foundational act of displacing or occluding the pregnant woman's body—whether by the monstrous body of a birthing male, or by a male-controlled birthing machine. Marianne DeKoven's memorable image for the "modernist vision of culture we customarily accept as inevitable" is a powerful articulation of this understanding: "[The] rough beast . . . its hour come round at last, slouching through the Waste Land . . . toward a scene of monstrous childbirth (made monstrous, of course, by the repression of the mother's body)."[43] A similar image of a monstrous birth, effecting the scientific and social marginalization of the mother's body, grounds the inaugural work in the feminist critique of science: Mary Shelley's *Frankenstein*.[44] Evelyn Fox Keller has anatomized modern science as a scopophilic project to penetrate the mother's body, grounded in male envy of female procreative power.[45] Both modernism and modern science, then, have been defined in and through the repression of the birthing woman's body. The female modernist writer's response to this act of repression is often a determination to reinscribe, in her writing, the embodied female subject. Yet such a project inevitably produces an anxiety-laden double bind, for the woman writer must rely on the doubled strategy of working in hegemonic discourse *while also* recuperating unwarranted, devalued aesthetic modes. Reinscribing the "forgotten vagina," women writers must tolerate the anxiety caused by the return of that which has been powerfully culturally repressed.[46] Charlotte Haldane's writings reflect these strategies, as well as the mingled anxiety and feminist determination they produce.

"THIS IS MY MAN": WRITING THE ROMANCE OF SCIENCE

In the first several decades of the twentieth century, a new understanding of the processes of conception and gestation, developed and promulgated by sexologists, embryologists, eugenicists, and naturalists, productively challenged conventional understandings of the boundaries of sex and sexual behavior. As Charlotte Haldane describes the origins of her first novel in her autobiography, *Truth Will Out*, she seems to parallel its intellectual gestation to that widely discussed biological process. And like its biological counterpart, this moment of literary conception blurs discursive, genre, and sex boundaries between literature and science: "It gave me an almost physical shock of excitement and pleasure, one of those mysterious shocks experienced by the creative artist at the first moment of impact of an 'inspiration' and . . . by the scientist also, when the solution of a problem or the outline of a new discovery similarly come to him" (15). Her image stresses the similarities between the (implicitly female) artist and the (explicitly masculine) scientist, but the narrative that follows figures not seamless union, but rather a contest between the two discourses for epistemological and ontological priority. After her artistic/scientific moment of inspiration, Charlotte realized that in order to write her novel she would "have to know a good deal about modern science, particularly biology" (16). She therefore began looking for "an expert, a scientist, preferably a biologist," to teach her the field. As with the moment of aesthetic conception, the story of her quest for a scientific adviser for her novel mingles the gendered categories of science and literature, objectivity and subjectivity, realism and romance, thereby destabilizing our understandings of modernism and modern science. It also reveals how Haldane's first novel was shaped by a characteristically modernist double bind: a surge of interest in the possibility of achieving control over reproduction and the discursive dilemma facing women writers who wished to exploit that new interest in reproductive control to advance feminist concerns.

In Charlotte's conception metaphor, beneath its assertion of the creative kinship of the artist and the scientist lies an unconscious sense of their rivalry. That potential for conflict colored her portrait of the developing relationship between Charlotte, novelist-to-be, and J.B.S. Haldane, the celebrated geneticist whose scientific advice she soon sought. In implicit acknowledgement of that rivalry, *Truth Will Out* carefully establishes the priority of her literary vision (her moment of inspiration for the novel) over his scientific counsel. She had already completed the outline of her projected novel, she reveals, when a colleague's chance loan of the condensed version of Haldane's *Daedalus* (published just six months earlier in *Century* magazine) set her "imagination aflame" (17). Charlotte marveled at the physical courage of this man who "specialised in making experiments on himself with some substance called acid sodium phosphate," and at his intellectual courage in "making startling predictions about the biological future of the human race" (16–17). But most impressive of all was his "fantastic but matter-of-fact account of the growing of a human foetus in the laboratory." "'This is my man!' I thought instantly, but wholly unaware of the personal implications this discovery was to have for me" (17).

J.B.S.'s image of an ectogenetic foetus embodied a fantasy of scientific control of reproduction like that fueling the series of articles on prenatal sex determination Charlotte had just published in the *Daily Express* and the *Sunday Express*.[47] Moreover, because he was a scientist, J.B.S. not only shared that interest but could legitimate it from his position of scientific authority. Yet there was a personal aspect to his appeal as well, revealed when Charlotte's realistic account of her search for a scientific adviser segues into the romance narrative of her search for "my man."

Despite its titular claim for unvarnished accuracy, Charlotte's narrative of their developing relationship in *Truth Will Out* relies on all the romance tropes of mistaken identity, glamorous history-laden surroundings, and forceful masculinity to figure her search for literary truth as a love affair—not just with the literary potential of modern science, but with the romantic potential of an individual modern scientist. The search for "my man" begins in oedipal confusion, when she searches for references to the author of *Daedalus* in the Oxford University Press library and finds not J.B.S. Haldane but his father, the humanist physiologist John Scott Haldane. After differentiating father (despite his "distinguished scientific publications . . . not the one I was after") from son (who initially appeared to have "no academic degrees"), she tracks J.B.S. to Cambridge, where he is a Reader in biochemistry. When he fails to answer her letters requesting an interview, "the urgency of [her] desire to get to work on [her] novel" propels her to the "earthly paradise" of Cambridge, where she accosts him in his "exquisitely proportioned" rooms (17–18). Meeting "my man," the scientist of her dreams, Charlotte is swept off her feet. She has found her "predestined teacher." Yet their relationship transcends the intellectual: "It soon became clear that my don and mentor was not interested merely in the cultivation of my mind." "With a charming affectation of eighteenth century gallantry" he implores her "favors," and she accepts his sexual desire for her because of her intellectual need for him. "Wholly in love with his mind," she recounts, she "did not withhold" her favors. "In any case, coyness was not natural to me" (20).

The memoirs Charlotte recorded for the BBC, late in her life, take up the story of their courtship still in the romance mode, constructing it according to the formula of love surmounting all obstacles. When she met J.B.S., Charlotte was already married, though unsatisfactorily, to an invalided veteran of World War I. Yet J.B.S. "insisted on marriage. The difficulties acted as a spur to his desires" (23). Therefore, the lovers had to engage in a documented act of adultery, because the legal system prohibited divorce on the grounds of simple desertion. As Charlotte recalled, Haldane "saw himself in a quixotic light, as the chivalrous rescuer of a little woman in need of protection." Although the Cambridge University Senate failed to share "this romantic view," resurrecting the aptly named "Sex Viri" committee to investigate

charges that a reader at the university had shown "immoral inten-
tions," J.B.S. was ultimately acquitted of the charges against him. The
story of their courtship ends, like all romance novels, with "the happy
union of the two principals."[48]

Just as Charlotte Haldane used the romance genre to structure the
account of her love affair with the scientist whose expert advice enabled her to write her first novel, so it was the romance genre to which
reviewers of that novel overwhelmingly responded. Reviewers seem to
have read *Man's World* primarily as a romance, and only secondarily as
a critique of science. The *Labour Magazine* called it "a prophecy, a romance, and . . . a protest of modern woman against opposition to her
ideals"; the *Aberdeen Post and Journal* praised the novel's "calm yet passionate love, born of intimate knowledge, of biology and sociology,
treated with the freedom of high romance." And in Charlotte Haldane's own paper, the *Daily News*, Gerald Gould, reviewing *Man's World*
under the heading "A Return to Romance," observed, "Mrs. Haldane
does not shirk shocking; but her heroine maintains the old and grand
ideal of romantic love." Even the publisher's publicity blurbs, which
were quite possibly written by Haldane herself, termed it "a romance
of the future."[49]

Does her reliance on the romance genre disqualify her works from
inclusion within the discursive categories of modernism or modern
science? As should be clear from my earlier discussion of those two
terms, I think not. In fact, Haldane's reliance on a feminized and devalued literary genre to narrate her negotiations with modern science
attests to her position within the newly mapped field of female modernism. As Suzanne Clark has observed, "Modernism created a rhetorical battlefield for women. As writers, they have had to rethink the
entire range of gendered relationships between authors and their
characters, between authors and readers" (*Sentimental Modernism*, 13).
And, we might add, between their choice of subject and the literary
community. As Clark points out, "The effects of modernism . . .
included not only denial but also recuperation."

Charlotte Haldane recuperated the romance genre because it permitted the introduction of a female subject and the experience of a
female body into her work. This subject position and embodied perspective on life made possible the articulation of a response to modern
science impossible in a work of art characterized by the masculinist detachment, objectivity, and scientificity characteristic of canonical male
modernism. Furthermore, just as J.B.S.'s scientific expertise recuperated (or authorized) Charlotte's interest in the broader cultural effect

of scientific innovations, as *Man's World* and her writing for the socialist magazine *Woman Today* both reveal, Charlotte Haldane embraced the romance genre in part because it could recuperate scientific themes for women readers.

MAN'S WORLD

Fundamental to the structure of *Man's World* is the double bind of female modernism. Charlotte Haldane adopts a romance plot centered on the (otherwise aesthetically unrepresentable) embodied female subject in order to confront *from a woman's perspective* the social implications of scientific control of reproduction. And she adopts the discourse of modern science, in particular its valorization of an empiricist perspective, in order to legitimate her main agenda: a celebration of female maternal agency. There are dramatic contradictions of effect between these two strategic positions. Casting her advocacy of female agency within the discourse of masculinist science, she seems to devalue gender-role atypical women, especially in relation to the scientifically authorized position of so-called normal motherhood. Using the trajectory of the romance to organize her narrative, she articulates female subjectivity at the price of trapping it within a love-and-marriage plot. But even in these strategic choices, "double-binding" as they are, Haldane's novel reveals the resistant doubling of meaning through which female modernist writers managed to find voice.

Man's World begins where Charlotte Haldane's essay "The Sex of Your Child" left off: with the assertion that discovery of a technique of prenatal sex selection could change the relations between sexes, nations, and races and solve what she understood as the problem of intersexuality. The novel's premise is the invention, by a geneticist named Professor Perrier, who is well versed in animal husbandry, of a system of exercises that enable prenatal sex selection. Designed to be performed regularly by the pregnant woman, the Perrier exercises enable the woman to choose a boy baby.

Man's World offers a prescient analysis of how reproductive technology can produce power/knowledge for a patriarchal state through control of the (female) body. Although designed to empower the pregnant woman, in reality the Perrier exercises trap her in a masculinist culture and ideology; directing her gestational empowerment toward the production of sons, they transform female choice into female necessity. They enable the state to choose the sex of each generation, and so use the scientific search for enlightenment to further its patriarchal, nationalist, and racist/colonialist mission.

Haldane's fictitious state in *Man's World* is dedicated to the perpetuation of the "entire white race," a task that recalls the theory of the survival of the germ-plasm (63). Promulgated by nineteenth-century German zoologist August Weismann, this theory held that a "particular sort of protoplasm . . . was transmitted substantially unchanged from generation to generation via the germ-cells, giving rise in each individual to the body-cells (soma) but itself remaining distinct and unaffected by the environment of the individual."[50] Kevles points out that this early articulation of the genetic basis of heredity was later embraced by eugenics groups concerned with inherited racial purity as an answer to those who pressed the influence of environment over heredity. As it was portrayed by biologists of the day, the theory was unconsciously, if not intentionally, gendered male, a gender bias Haldane's dystopia exploits fully.[51]

Haldane's novel not only portrays the sexism integral to eugenics in the late 1920s, it also anticipates the rise of Nazi eugenics within the decade. In 1933, the counselor of Germany's Reich Ministry of the Interior would justify the Eugenic Sterilization Law passed by Adolf Hitler's cabinet by evoking the purity of the racial bloodline: "We want to prevent . . . poisoning the entire bloodstream of the race."[52] A similar mythology of the purity of the blood structures the dystopian state of *Man's World*. Modeled on the human cell, its central city is named Nucleus, and its official propagandists have "translated the terms of the social organization into those of the human body. The Body then [stands] symbolically for the entire white race" (63).

Anticipating the abuses of the Nazi doctors, Haldane's fictitious state uses biomedical science to maintain racial division and white supremacy. The state relies on a worldwide network of "communication and direction"; a surveillance and control group called the "Ears," "founded strictly on the principles laid down by their psychopathological researchers"; and a technique for race-specific chemical warfare known as Thanatil, targeted against "that enzyme which produces the black pigment in negroes, and which, when attacking the tyrosine ester of Thanatil absorbed by the dusky skin, gradually liberates the poison till the central nervous system is invaded, causing paralysis and death" (63–65).

The fact that in Haldane's novel oppressive political uses are found for the Perrier exercises may seem to provide fictional confirmation for contemporary critics of reproductive technology, who hold that gendered objectification is integral to the scientific method.[53] Although the Perrier exercises testify to woman's power (for the woman's physical work produces the desired-sex foetus), once governments realize the method's potential to consolidate "Man Power," the prenatal production of male foetuses takes top priority (36). Boys are needed, the narrator explains, to perpetuate patriarchy, the patrilineal class system, and industry.

Indeed, the disproportionate value placed on masculinity in Haldane's dystopia does translate into a biologically based and objectifying vision of gender roles. Women are divided into three categories according to their reproductive activity. Vocational mothers are selected by state committee to participate in "a career which had its grades like all others" (55). They devote their lives to "the theory, as well as the practice, of race-production" (55).[54] "Neuters" occupy themselves with the professions, and "entertainers" serve men sexually and aesthetically—as dancers, actors, singers, poets and novelists—and must smile "perpetually" (130). These categories are rigid and impermeable. Women must choose their category at puberty, and at that point the other two are permanently closed to them by the intervention of state-enforced science: "Either you become a mother or you must be immunized" (127).

Yet resistance to such globalized reproductive control is still possible. Haldane's novel chronicles two tales of such individual, body-centered resistance to the compulsory reproductive categories enforced by this scientific state. There is Christopher, whose mother, mourning the loss of a daughter born "abnormal," refuses to practice the Perrier exercises during his gestation (296). As a result, he is born "intermediate sexually" (296). Resisting conscription into the ranks of the professionally and biologically "normal" males, Christopher prefers to be a musician and philosopher rather than a scientist, and to remain celibate rather than mate with an appropriate female partner. The second resister is Christopher's sister, Nicolette, who refuses to choose between motherhood and the two other socially enforced roles for women, Neuter professional and Entertainer. Instead, with her brother's help, she procures an antidote to the state-enforced sterilizing "immunization," planning to become pregnant *not* according to state policy but rather by her own free actions.

Both attempts at resistance fail, not because they are defeated from without but because they collapse from within. Haldane's representation of the power over human bodies produced by her fictional reproductive technology reveals that the very terms within which resistance arises may transmute it into an effect of the power it seeks to dislodge. Christopher's opposition to heterosexual normality is hollowed out and possessed by the normalizing discourse it opposes. With fatal consequences, he internalizes the restrictive categories of Nucleus, which label some people normal and others deviant, depending on their sexual and reproductive behavior. He comes to see himself in the terms of the dominant society: as one whose "submasculinity" prevents him from contributing to the improvement of the race (297). Unable to find support for his beliefs either from Nicolette or within himself, Christopher commits suicide by flying too high in his airplane. He crashes for lack of oxygen, a Daedalus turned Icarus.

Nicolette's resistance is directed not at gender roles but at sexuality, defined narrowly as reproduction. The state controls women's reproductive lives through the Motherhood Council, which assesses the women's fitness to be mothers in terms of their genetic makeup, character and education, and assigns reproductive partners—"mates"—to the women permitted to reproduce. Nicolette resists this state regulation of motherhood, arguing that it reduces female liberty, and chooses instead to regulate her own reproductive life.

Freedom is an elusive condition in Haldane's biologically deterministic society. Although Nicolette's resistance does not end tragically—as does Christopher's—she suffers a kind of death, for she is co-opted by the patriarchal and instrumentalist values of Man's World. She falls in love with Bruce Wayland, chief experimental scientist of Nucleus, and her resistance is transformed to loyalty. Pregnant by her scientist-lover, who calls her his little "mother-pot," Nicolette comes to think of herself as but an instrument for producing "his" son (295). High-placed in government, Bruce is able to recast her resistance to the state's mutually exclusive categories for female behavior (motherhood or sexual experience or a profession) as submission to the state ethos of auto-experimentation: "an experiment, although . . . unusual and a bit risky" (239). Accepting the romantic/reproductive/scientific contract at last, Nicolette both objectifies herself and is ready to accept objectification by others.[55]

Thus, power produces resistance that turns into power—both Nicolette's socially constructed, limited power as a mother-to-be and the patrilineal power soon to be enjoyed by the archetypal masculine subject who will be (re)born from her womb. Pregnant, Nicolette affirms not just experimental science but also the specifically *vocational* motherhood she previously resisted, whose central concern is the creation of a son through the Perrier technique of prenatal sex selection (250–251). The novel only implicitly acknowledges the crucial fact: that Nicolette's sort of power is fundamentally different in kind and in degree from that possessed either by Bruce or by their son-to-be.

To argue that Haldane's novel embodies the oppressively gendered implications of prenatal sex selection is not to say that Haldane consciously planned to write a dystopian novel. Rather, the novel may have been intended, and was certainly received, as a feminist utopia. Critics praised "the wife of the well-known Cambridge biologist" for enriching "the literature of Utopia," while ironically emphasizing the novel's feminist agenda, as "a protest of modern woman against opposition to her ideals."[56] But the novel betrays at least an ambivalent response to the scientific project, nowhere more vividly than in the character of scientist Bruce Wayland.

Bruce is modeled on J.B.S. Haldane and explicitly associated with the two scientific activities that first attracted Charlotte Haldane to her husband to be: auto-experimentation and ectogenesis (see *Truth Will Out*, 16–17). Yet despite this initial attraction to J.B.S., and the ectogenesis and auto-experimentation with which he was associated, *Man's World* dramatizes the oppressive foundations of these scientific activities. Two scenes particularly merit a closer look, because they dramatize the frightening side of Bruce Wayland and his scientific practice: an early scene in which Wayland defends ectogenesis to the vocational mothers of Nucleus and a later scene in which he protests his ban from the laboratory where he engages in auto-experimentation.

In the first scene, in which the vocational mothers of Nucleus discuss ectogenesis with Bruce Wayland and a visiting geneticist, Charlotte Haldane dramatizes her fear that the new technology will displace women from their reproductive (and social) roles.[57] The geneticist, who developed the technique in his work on cattle, asks the mothers how they think "the suggestion of human ectogenesis will be generally received?" (59). The response he receives is unequivocally negative: "You will be the most unpopular man in the world" (59). Yet despite the women's horror at the notion of "a sort of human termite queen . . . from whom the entire race shall be bred," Wayland joins the geneticist in impassioned defense of the reproductive technology on eugenic grounds: "Ectogenesis provides the means to select on the most strictly accurate lines. The numbers of mothers chosen diminish year by year. Until at last, those who supply the race are the supreme female types humanity can produce" (61–62). Claiming a stance of scientific objectivity not yet shaken by the revelations of Nazi bio-medicine, Haldane here shows a frightening indifference to the re-duction of women to breeders.[58]

Haldane continues her attack on scientific objectification in the second scene, figuring a debate between Wayland and the Company Director, who bans Bruce from further auto-experimentation. The very parameters of this debate express the limitations of instrumental rationality: either Bruce can do what he wants with his body, and so he can continue auto-experimentation, or the Company Director can do what the company wants with Bruce's body, and so Bruce cannot con-tinue with his dangerous auto-experimentation. But no matter which form of instrumentality prevails—scientific or industrial—the out-come is the same: Bruce's body is objectified and alienated, con-structed as something to *use* rather than something to *be*.

If we tease out the implications of auto-experimentation and ecto-genesis as embodied by Bruce Wayland, we discover that both activities express an ideology privileging scientific instrumentality or, as Bruce puts it, viewing "all living and striving . . . [as] amenable to experi-ment" (62). Both procedures embody notions characteristic of West-ern post-Enlightenment rationality and shared by both industrial and reproductive technology: a mind/body split that valorizes mental ex-perience and denigrates physical experience and a notion of the au-tonomous individual that, coupled with the notion of the body as property, has been marshaled to support acts of bodily objectification as diverse as prostitution, organ-selling, and surrogate motherhood. In short, both procedures supply the foundation for a critique of in-strumental science.

Yet Charlotte Haldane's pseudoscientific utopian novel falls short of the critique of science I have sketched out above. This happens in part because *Man's World* relies on scientific discourse to advance the cause of female agency and autonomy. But there is another reason why Haldane's novel eludes the clear-cut criticism of the scientific project that contemporary feminist readers might desire: her position as a woman writer working in a literary field all too often constructed as *man's world* as well. The ambivalence toward science betrayed by Hal-dane's novel is rooted not only in Haldane's personal experience but also in her position as a woman writer in relation to literary modern-ism. *Man's World* embodies the characteristic traits of female modernist writing: *attention to the experience of marginality*, in its concern with the problem of the intersex; *concern with gender politics*, in its attention to the gendered implications of such reproductive technologies as ectogenesis and prenatal sex selection; *strategic use of a decentered perspective,* in the use of two protagonists of opposite sex (Christopher and Nicolette), one of whom resists, one of whom capitulates; and *the split focus or doubled gaze resulting from conflicted identification,* in the ambivalent representation of scientific instrumentality as a force both sexually appealing (to Nicolette) and life-threatening (to Christopher).[59]

To those structural and thematic traits of female modernist writing, we can add the fifth trait shared by late-Victorian and modernist woman writers: *the (strategic or involuntary) use of scientific language to advance the feminist cause of female agency and autonomy.* Haldane's novel reflects this trait in its theme, characterizations, and plotting, all of which are part of her intimate negotiation with the hegemonic discourse of science in order to achieve the conditions for expressing her feminist, socially informed vision.

Tracing the development of the scientific state from control over reproduction (via birth control, then sex predetermination) to control over women, *Man's World* ends with Bruce's chilling assertion that "there will always be Christophers, and they will always suffer. But it's the experiment that counts for us, not the result" (299). Yet Haldane has taken care to show us, and make us care, not just about her characters' experiments with maternal agency and sex determination but about their results. We care both that Christopher is forced to suicide by a society unable to tolerate his intersexuality and that Nicolette lives, achieving motherhood on her own terms and thus expanding the social possibilities for women. Poised between celebration and critique of the scientific control project, *Man's World* is—to contemporary feminist readers who respond to its prescient political analysis—a profoundly troubling dys/utopia that vividly exemplifies the double bind of female modernism.[60]

FROM *MAN'S WORLD* TO *WOMAN TODAY*

"Happy ever after" is a fairy-tale ending, whether the love affair is with a man or with a way of seeing the world. At approximately the same time that she fell out of love with J.B.S., Charlotte Haldane gradually became disillusioned with the analytic power of science. As she recalled in her autobiography:

Having absorbed as much as my untrained mind could master of the scientific outlook on life, I began, at first slowly, with emotional resistance to admitting my disillusionment, but gradually more rapidly, to

> lose my interest in science. This coincided with my realisation that my
> second marriage was not going to give me the satisfactions, especially
> the children, which I had hoped for from it. So I began to look for in-
> tellectual and emotional compensation in other directions. I turned
> back to literature, and to the more familiar domain of the written
> word. (*Truth Will Out*, 33)

But before she broke decisively either with J.B.S. Haldane or his scientific worldview, she experimented one last time with mobilizing for her
own political purposes the privileged access to scientific discourse that
her marriage provided. Here too, in her writing for the socialist
monthly *Woman Today*, she revealed the qualities that characterize female modernism. But this time her goals were a potent and conflicted
mix of socialism, antifascism, and feminism.[61]

In 1939, Haldane took on the editorship of *Woman Today*, a paper
published by the Women's Committee for Peace and Democracy.
Woman Today had a monthly circulation of more than 2,500, mostly
sold through Left Bookshops, and boasted the support, among others,
of MP Ellen Wilkinson and novelists Rosamund Lehmann and Rebecca West.[62] The magazine printed a steady stream of leftist fiction
and features, including Sylvia Townsend Warner's series "Women of
Yesterday" (Harriet Beecher Stowe, Rosa Luxemburg, Countess
Markievicz, and Josephine Butler), short stories by Naomi Mitchison,
and monthly editorial essays discussing topics ranging from women's
role in the Spanish Civil War to the Chinese women's movement.[63] Reflecting Haldane's long-standing interest in the area, the magazine
also published essays on scientific topics, such as Dr. Barbara Holmes's
discussion of the early forays into estrogen replacement therapy, "The
Gland That Controls Your Sex."[64] As this brief survey suggests, the
magazine combined the domestic discourse of the woman's magazine
and the authoritative discourse of science to advance the causes of antifascism and socialism.

An article by Haldane herself perhaps best illustrates how the magazine used the discourse of science for left-wing political purposes,
often muting its feminism in the process. Titled, "'They Were Two
Hours from Death, But I Was Not Afraid': The Inside Story of My
Husband's Experiment," the piece records how J.B.S. Haldane and
four other members of the International Brigade experimented on
themselves to further scientific knowledge, and ultimately to determine why the British submariners died in an accident on the subma-
rine *Thetis*.[65]

The essay recycles themes familiar from *Man's World*—the bravery of auto-experimentation, the social centrality of science—but with a crucial difference of emphasis reflecting its socialist commitment. The narrative strategies of the woman's magazine sugarcoat her subject, producing an idealized portrait that subordinates critique to a celebration of the scientist's authority. Thus Haldane begins with an appeal to the curious lay reader: she promises to reveal "what it feels like to be married to a scientist who occasionally experiments on his own body to find out things for the benefit of humanity" (2).

If the feminist critique of science has dropped out, however, the revisionary leftist critique has replaced it. In an ironic return to—and deconstruction of—the notion of enlightened government by scientists central to *Man's World*, Haldane praises the "scientific tradition" as "one of the noblest conventions of mankind," while denying that a link exists between science and the modern post-Enlightenment state (3). Charlotte Haldane constructs science not as gendered oppressor, but as ungendered site of resistance. She shows scientists working not to consolidate (masculinist) state power, but to reduce the human abuses (industrial and military accidents, illnesses) produced by capitalism. Her break with J.B.S. and his scientific worldview obviously imminent, she rejects the glorification of auto-experimentation as the pinnacle of human self-sacrifice and courage, exalting instead the greater courage of ordinary citizens. Implicitly rejecting the notion of a scientific elite controlling a debased and passive citizenry, she puts aside the specifically feminist analysis of her earlier critiques of science, instead urging "the common men and women of this country" to work together with scientists "to overthrow this system and to bring in Socialism, Peace, true Democracy, and a really Brave New World" (3).

From "The Sex of Your Child" to "a really Brave New World," Charlotte Haldane's writings seem a puzzling mix of complex, even frustratingly inconsistent, ways of defining and responding to social injustice. In *Man's World*, she interwove a scientifically articulated endorsement of female reproductive and social self-determination with a critique of the objectifying character of empiricist science; in *Motherhood and Its Enemies* she collaborated with the scientific construction of woman-as-mother in order to combat the greater eugenic threat to woman's maternal agency. Finally, as editor of *Woman Today*, Haldane mobilized the rhetoric of conventional wife- and motherhood to leftist, antifascist ends. (Ill. 16.)

How do Charlotte Haldane's writings, from the 1920s through the 1940s, on prenatal sex determination, intersexuality, and motherhood

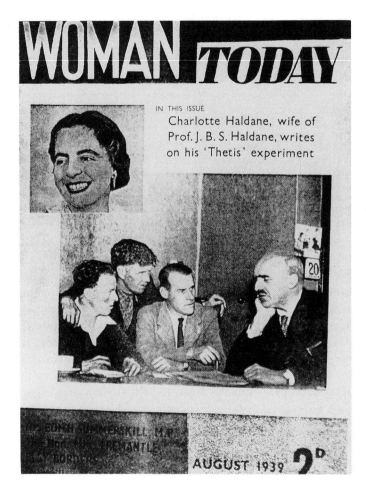

16 Charlotte and J.B.S.

illuminate our contemporary response to, and representation of, reproductive technology? If we read the contradictions in her works symptomatically, we can understand them as reflecting her complex position as a woman within literary modernism. They reveal how a modern woman's response to, and representation of, those technologies can be shaped by her particular social positioning. Thus women writers who wished to engage with the social and psychological implications of modern science were forced to adopt compositional strategies that seemed anything but modernist.

Charlotte Haldane chose a form far from high modernist for her novel of prenatal sex determination for precisely such strategic reasons. When she first read J.B.S. Haldane's account of "growing a human foetus in the laboratory," she was sure she had found the ideal adviser for her novel, and she expressed her sense of instantaneous affinity in the discourse of romance fiction: "This is my man!" Her attraction to ectogenesis, like her choice of the romance genre to express it, took the particular form it did because of her position as a woman writer. As I discussed earlier, feminist scholars have revealed as the fundamental impetus to both modernism and modern science a process of repression and denigration of the birthing woman's body. For J.B.S. Haldane, Julian Huxley, and—as we shall see—Aldous Huxley, this process was powerfully expressed through the image of ectogenesis. Yet if for male writers in the modern period (whether identified with canonical modernism or not), the image of ectogenesis fed a fantasy of male agency and autonomy, acquired through denigration of the image of the gestating woman, for women writers the image meant something rather different.

Charlotte Haldane saw implications in ectogenesis diametrically opposed to those envisioned by her male friends and colleagues. Her goal as a modern woman writer was to empower women, particularly mothers. Thus her strategy as a writer was to recuperate, and reintroduce into literary consideration, the birthing woman *as embodied subject.* Since she wanted to write a novel extrapolating from the scientific control of reproduction to its social implications, she seized on J.B.S. as the ideal source of information and legitimation for *her* representation of science. She then used his scientific prestige as her warrant to consider science and its achievements not in a social vacuum, but in relation to human society. In short, Charlotte Haldane used J.B.S.'s scientific prestige to advocate a position not scientific but social: greater agency and autonomy for women, as mothers.

If, as I have argued in a previous chapter, a symptomatic reading of J.B.S.'s interest in ectogenesis views it as a strong male defense against the anxiety raised by reproductive asymmetry, a symptomatic reading of Charlotte's interest in prenatal sex selection would view it as a strong female strategy for combating gender-based social asymmetry. Working through science, focusing too on the pregnant or birthing female body, Charlotte Haldane's writings are a particularly charged site for negotiating the conflicted relationship between women writers and the scientific, social, and literary milieu of modern Britain.

What does the example of Charlotte Haldane's fiction and nonfiction have to say about the relation of women writers to literary modernism? The position that her works articulate on the scientific issues of the day reflects her complex and conflicted situation as a modern woman writer for whom agency, subjectivity, and embodiment are all deeply problematic. Her writing exemplifies the strategic doubleness, as well as the double bind, of female modernism. Trapped within the discourse of science, she wielded it subversively to release, rather than to repress, female agency. Limited to the genre of the romance novel if she wanted to adopt a woman's voice and perspective, she used the romance plot as the platform for creating an embodied female subject who could transcend it.

Charlotte Haldane's works exemplify the unacknowledged social and aesthetic questions shaping the debate about prenatal sex determination, intersexuality, and motherhood from the 1920s through the 1940s. Moreover, they suggest that uninterrogated conflicts and double binds resulting from our social and discursive positioning as women in modern (and, now, postmodern) culture may structure our experience of the debate over reproductive technology, as well as the positions we choose to argue. The example of Charlotte Haldane reminds those of us working in the modern period to resist the tendency to assume a congruence between discursive style, gender identity, and ideological position. As her writings in the early years of this century attest, resistance to reproductive control can appear in unexpected places. A modern feminist can have many strategic reasons for embracing the discourse of modern science or the woman-centered arc of a romance plot.

four

Embryos Are Like Photographic Film

RTS AND VTS IN THE FICTION OF ALDOUS HUXLEY

Visualizing and imaging technologies are critical to the technical and discursive apparatus of assisted reproduction.
—SARAH FRANKLIN, "Postmodern Procreation"

It is possible to make use of the movies to improve our vision for objects and events in real life.
—ALDOUS HUXLEY, *The Art of Seeing*

CONTEMPORARY CONTEXT: THE VT/RT LINKAGE

What do the Hubble telescope and the scanning tunnel microscope have to do with embryo research? A quick answer to that question would be to say that all three are associated with high technology and contemporary science. Perhaps that was why Sir Ian Lloyd—surprisingly—invoked those scientific instruments in his discussion of embryo research, during the 1990 House of Commons debate over the Human Fertilisation and Embryology Bill:

> As science takes us nearer to the fundamentals of creation, whether through the outward reach of the Hubble telescope which is to be launched into space next week, or the inward reach of the scanning tunnel microscope, revealing for the first time the secrets of the living cell, we shall be presented with greater potential for good and evil, greater powers of intervention, and greater challenges to orthodox dogmas of all kinds—religious, scientific and political.[1]

Yet it is not enough just to say that these instruments are connected to embryo research by their status as "high tech" science. Although accurate as far as it goes, that explanation does not exhaust Lloyd's reasons for mentioning them in such a context. Rather, his choice of the advanced microscope and telescope to exemplify the implications of embryo research reflects a long-lasting, overdetermined association between visualization technologies (VT) and reproductive technologies (RT).

Many contemporary feminist theorists hold this VT/RT link responsible for a dramatically new phenomenon: the contemporary construction of the gestating woman as an invisible, even hostile environment for the fetus-at-risk.[2] Beginning with the crucial work of Rosalind Petchesky, those theorists have argued that such (relatively) new medical visualization technologies as ultrasound and fiber-optic endoscopy have enabled the medical separation of the fetus from the gestating mother, permitting the [re]conceptualization of the fetus as not a part of, but rather apart from, the woman who is gestating it.[3] The fetus thus becomes available for legal and medical interventions on its own behalf, as if it were already a civil subject.

Although I share others' alarm at this postmodern maternal/fetal disjunction, I do not share the general opinion that the VT/RT link it reflects is a new phenomenon. Rather, I argue that this representational tendency to split the fetus from the mother-to-be reflects the prominent role played by visualization technologies in the development of reproductive technology. A VT/RT link has been written into the history of reproductive technology from its earliest years—as exemplified by its prefigurative conception in the writings of Aldous Huxley. The image of fetal autonomy, the strategic separation of gestating mother from fetus, is merely one among several disturbing results of that linkage. Recovery of the history of relations between visualization and reproductive technologies can both illuminate how the developing fetus is portrayed in contemporary representation and enable us to develop more flexible responses to such representations.

"FROWARD HOMUNCULUS"

"If I had been able to go through with the biological and medical education, which was interrupted in my youth by a period of near blindness," Aldous Huxley wrote a friend in 1949, "this is what I shd [sic] have liked to become—a fully qualified striker at the joints between the separate armour-plates of organized knowledge. But fate decreed otherwise, and I have had to be content to be an essayist, disguised from time to time as a novelist."[4] The fate that kept Aldous Huxley from medical education was an onset of *keratitis punctata*, a corneal inflammation caused by *staphylococcus aureas* contracted while he was at Eton.[5] This deflection from his earliest ambition—to study biology and medicine—was only temporary, however. In his adult writings, Aldous Huxley would return to those themes with the particular inter-disciplinary twist and visual emphasis determined by the episode that temporarily deflected him.

From his earliest years, Aldous leaned toward scientific pursuits, re-flecting influences we can briefly summarize: the writings of his grandfather T. H. Huxley, naturalist and defender of Darwin's evolu-tionary theory, a particularly dominant family presence during Al-dous's early years because his father, Leonard Huxley, was writing the *Life and Letters of Thomas Henry Huxley* (1900); the influence of his elder brother Julian, who presented him with his first encyclopedia ("I now know everything," Aldous wrote in thanks, "from who invented dice to the normal temperature of the sea-cucumber"); his years at Hillside preparatory school where his "lasting interest in natural history was stimulated by the system of 'notices' awarded by the tough veteran master, Mr Taylor, to the boys who each summer recorded the first specimen of different species of butterflies and moths"; Eton, which he entered already intending to specialize in biology, and where his scientific interest was further kindled by the science master M. D. ("Piggy") Hill; and not least the feverishly scientific atmosphere of Cherwell, the family home of John Scott Haldane, where Aldous boarded during his time at Oxford in the war years.[6] There, he dis-cussed literature with Naomi Haldane. As she later remembered, "While I was learning from him he was picking up science and a particular attitude towards it from my father, who is in a sense the original of Lord Edward [Tantamount] in *Point Counter Point*.[7]

However, Aldous's hopes of becoming a biologist or doctor were rudely dashed in 1911, while he was still a student at Eton. The eye infection he contracted, only finally diagnosed more than six months later, left him nearly blind at the age of sixteen. Although his sight did eventually recover enough to permit him to read with a powerful mag-nifying glass, the illness first subjected him to eighteen months of near darkness, during which with remarkable stoicism he taught himself to read Braille, play the piano, type, and even to ride the bicycle again. Although he seldom spoke of the ordeal of his blindness, or of the con-tinuing vision problems that ensued, "the art of seeing" became a cru-cial theme in Huxley's fiction, even ultimately the topic of a nonfiction work of the same name. Unable to engage in the scientific explorations open to a biologist or a medical doctor, Aldous instead would write about the impact of science on society: science conceptualized, metaphorically and pragmatically, as the struggle to see nature clearly.[8]

During his college years of near blindness, living with the Haldanes at Cherwell, Aldous mapped out a plan for scientifically informed literary studies: "One might tag on to literature-study a sketched history of thought—old conceptions of medicine, development of alchemy and astrology into scientific channels . . . while, possibly, one might touch on the development of religion and philosophy."[9] Five years later, he published a poem that turned the history of science to literary ends, introducing the specific scientific theme which would figure prominently throughout Huxley's writing life: sexual reproduction. "Fifth Philosopher's Song" (1920) is an early illustration of the RT/VT link central not only to Huxley's work but also to our twentieth-century representation of reproduction.

FIFTH PHILOSOPHER'S SONG

A million million spermatozoa,
    All of them alive;
Out of their cataclysm but one poor Noah
    Dare hope to survive.

And among that billion minus one
    Might have chanced to be
Shakespeare, another Newton, a new Donne—
    But the One was Me.

Shame to have ousted your betters thus,
    Taking ark while the others remained outside!
Better for all of us, froward Homunculus,
    If you'd quietly died![10]

Representing pregnancy as the product of spermatic/paternal agency, with the mother acting only as shelter for the developing homunculus, Huxley invokes an understanding of pregnancy reaching back from the alchemists to the Greeks. In its reproachful address to the "froward Homunculus" whose life has ousted another potential Shakespeare, Newton, or Donne, the poem alludes to the understanding of conception advanced by the Renaissance alchemist Paracelsus. By Paracelsus's account, the fetus originated in "a true and living infant, having all the members of a child that is born of a woman, but much smaller. This we call a homunculus."[11] As Huxley's older contemporary Joseph Needham would point out in his monumental *History of Embryology*, Aristotle promulgated the theory that semen provided the form for the embryo and the female blood supplied the matter to be shaped. The Aristotelian construction of the relative parental contributions (*sperma* and *catamenia*) was markedly asymmetrical; the formative importance of the paternal role, to Needham, recalled Aeschylus's *Eumenides*: "The mother of what is called her child is no parent of it, but nurse only of the young life that is sown in her. The parent is the male, and she but a stranger, a friend, who, if fate spares his plant, preserves it till it puts forth."[12]

"Sex, as the most striking and piquant instance of the universal irony, concerns [Huxley] much," quipped a reviewer of "Fifth Philosopher's Song."[13] But the reviewer got it wrong: what Huxley's poem represents is not sexual intercourse, but reproduction. Bodies engaged in sexual congress are nowhere to be seen, nor is the body of the gestating woman. Only the spermatozoa, metonymic representatives of the speaker himself, have any material existence. (Even that, of course, is challenged ironically.) In the modern world of Huxley's poem, sexuality and reproduction have been separated, and social improvement ("Better for all of us") is now being negotiated through reproductive intervention. The "froward Homunculus" is rebuked for his selfishness in attaining the ova: "Better for all of us . . . if you'd quietly died."

Huxley's modernist meditation on the vagaries of chance and the origins of the individual links scientific and religious discourses to celebrate (albeit ironically) a biological event: the moment of conception. The poem derives its satiric effect partly from its unexpected juxtapositions in space and time; its apostrophe to the sperm and the "froward Homunculus" (with both of which the speaker explicitly identifies) and its move from celebrating the possible plenitude of creation to the specific survival of "The One that was Me." And central to Huxley's representation of reproduction—although not as yet thematized—is microscopic vision. Those "million million spermatozoa" trigger his imaginative meditation on the philosophical problems of individual agency and collective good. As H. J. Muller might have put it, the poem assumes that the "microcosm" is key to the "macrocosm."[14]

The link between reproduction and vision that is expressed through microscopic images in "Fifth Philosopher's Song" broadens, throughout Huxley's fictions of the 1920s to include the preeminent modern visualization technology: the cinema. In Huxley's savage social satire *Crome Yellow* (1921), Scogan—like Ian Lloyd after him— makes a connection between reproductive and visualization technologies, RT and VT. In a devastatingly ironic speech (thought by T. S. Eliot to have been lifted verbatim from Bertrand Russell's conversations during Ottoline Morrell's celebrated Garsington house parties), Scogan connects methods of biological reproduction to what Walter Benjamin has called a technique of mechanical reproduction: the cinema.[15]

> With the gramophone, the cinema, and the automatic pistol, the goddess of Applied Science has presented the world with another gift, more precious even than these—the means of dissociating love from propagation. . . . In the course of the next few centuries, who knows? the world may see a more complete severance. I look forward to it optimistically. Where the great Erasmus Darwin and Miss Anna Seward, Swan of Lichfield, experimented—and, for all their scientific ardour, failed—our descendants will experiment and succeed. An impersonal generation will take the place of Nature's hideous system. In vast state incubators, rows upon rows of gravid bottles will supply the world with the population it requires.[16]

Huxley's image in this passage provides a bit of historical background to the maternal/fetal disjunction that many feminists currently understand to be a product of the latest medical imaging methods. However, no dramatic technological shift toward postmodern visualization technologies is responsible for this severance of sex from reproduction, fetuses from gestating women. Instead, Huxley explicitly constructs three modernist innovations as preparing the way for a fourth, the still hypothetical reproductive technology of extrauterine gestation, ectogenesis. Just as the automatic pistol, the gramophone, and the modernist cinema function through dissociation (of the firing mechanism from the operator of the gun, of sound from instrumentation, of image from context of image), so too this image of ectogenesis splits the gestating mother from the fetus and, in the image of "gravid bottles," does away with woman's body altogether.

Although the RT/VT link that Huxley makes in *Crome Yellow* is echoed in similar passages in *Antic Hay* (1923), *Point Counter Point* (1928), and *After Many a Summer Dies the Swan* (1939), *Brave New World* (1932) and *Ape and Essence* (1949) most explicitly focus on the relations between human reproduction and the cinema, while privileging different poles of the linkage. We can begin our consideration of Huxley's representation of human reproduction by assessing the meaning, for him and his contemporaries, of that dissociative modern visualization technology: the cinema.

CINEMATIC PRINCIPLES AND REPRODUCTIVE SCIENCE

In 1929, Aldous Huxley's good friend Robert Nichols contributed the essay "The Movies as Medium" to a journal on whose editorial board Huxley served. This essay, published in *The Realist: A Journal of Scientific Humanism*, can serve as a point of departure for a consideration of Aldous Huxley's links to, and understanding of, the film community.[17] Although Nichols's essay begins with some snipes at Stravinsky and the Imagists that have not worn well, his insights into film theory are more enduring and suggestive. Nichols formulates three principles for the essence of movies as a medium: (1) "Visual flow unlimited by the physical is what differentiates the medium [film] from all other arts"; (2) "The soul of the motion picture medium is the camera (under which term I include all the mechanics of receiving, retaining, and projecting the visual flow), because through it and it only can What-we-will be created"; (3) "In what we present to the camera and how we make the camera receive it lies the art of the motion picture. . . . For the peculiarity and beauty of the camera is that it can lie and keep on lying."[18] We might summarize Nichols's understanding of the essentials of film as follows: movement, the satisfaction of desire through the mechanics of the camera, and the camera *lie*.[19]

Movement is crucial to the cinema not just as an aesthetic element but as both a scientific instrument and the subject of scientific investigation. Eadweard Muybridge used his "zoopraxiscope" in the 1880s to illustrate his lectures on animal locomotion.[20] Etienne-Jules Marey's invention, the *chronophotographe*, "a synthesis of photography and techniques for recording space," was adapted by Félix-Louis Regnault and Charles Compte in 1895 for "a series of time motion studies considered by many to be the earliest example of ethnographic film," according to Fatimah Tobing Rony.[21] And in that same year, the cinematographe of Auguste and Louis Lumière spawned a range of documentaries whose most dominant form was "the simple recording of reality, such as workers leaving a factory."[22]

What these three examples from the early history of film technology share is the strategy of using filmic recording of motion as a scientific, taxonomic device. Motion is recorded because it enables the viewer/recorder to categorize, study, and understand the moving object, be it a member of another species, another race, or another class. The applications of such filmic recordings of motion were both military and industrial. Etienne Marey's "photographic gun," invented in 1881, so impressed the French military authorities, according to Doray, that they allowed him use of the Station Physiologique at the Parc des Princes, featuring "a 500-metre circular running track, mobile platforms for photographers and large black backdrops" (*From Taylorism to Fordism*, 79). Moreover, they also had an industrial application, Doray explains: "These photographic methods prepared the way for the calibration and standardization of movement. They were later to be adopted by Taylor's associates, and they are still used today in some work-study departments" (80). That the time and motion studies of early-twentieth-century photographic pioneers were picked up by Frederick W. Taylor and his followers attests to the ultimate motive beneath such cinematic recordings of motion: the desire to control human motion toward industrially efficient ends.

By the turn of the twentieth century, the taxonomic function of motion had taken on zoological and eugenic overtones. Two films made by Aldous's older brother, Julian Huxley, in the decade following Robert Nichols's pioneering bit of film criticism, can illustrate how the cinematic study of motion was used to gratify desire in both scientific and mass-culture contexts. The context for these films is the rise of the British documentary film industry, which reflects the increasing interest in popular science writing, as well as the more general increase in public relations as an agent of the modern democratic state.[23]

The Associated Realist Film Producers, an organization of documentary film producers, brought together a group of scientists "active in arguments over the social relations of science"; Julian Huxley and J.B.S. Haldane were prominent members.[24] Julian Huxley was a central figure in the documentary film community in London during the 1930s, making more films than any of the other 1930s scientists involved in the documentary film industry, and very visible in film organizations. As Tim Boon summarizes them, Huxley's projects ranged from state-funded documentary to leftist propaganda:

> He was a founder member in 1925 of The Film Society, an organisation devoted to showing films denied certificates, or otherwise little seen. In December 1935 he became one of nine consultants to "Associated Realist Film Producers," an organisation established to enable

members of the documentary movement to ply for trade outside the GPO [General Post Office], institutional home of the Movement from 1933. Further, he was one of several liberal figures on the unlikely General Council of Kino, the film distributors associated with the Communist Party.[25]

The diversity in Huxley's film work extended to the films he made himself, which reflect a range of genres characteristic of the medium, in particular in its scientific application. There were two dominant non-feature genres for 1930s propaganda film according to Boon, both of which were used, to some degree and at different times, for scientific or eugenic purposes: instructional films, which were didactic and at least supposedly ideologically neutral in tone, and documentary films, which "stressed the social function of the film as a tool of democracy."[26] But the notion of using film to instruct or convince on scientific issues was not limited to the nonfiction genres. Certain sentimental, fictional genres also appealed as vehicles for scientific information or exhortation, as an early and abortive foray into melodramas by the Eugenics Society reveals. A draft scenario for a possible film, sent in 1919 by the International Cinematograph Corporation to the Eugenics Society, suggests how melodramatic situations could be constructed to point an eugenic moral: "'Vasectomy, A Drama dealing with Eugenics' in which the advice of a eugenically-inclined doctor recommending vasectomy allows a wealthy but eugenically unsound couple, Walter Weakly and Lily Tremblett, to marry and adopt the 'numerous family' of the impoverished but eugenically-sound Paul Power and Rose Boniface."[27] The Eugenics Society moved away from melodrama to the instructional film, which as Boon points out, was "a form associated strongly with scientific subjects," rather than with political and social commitments.[28] As I will demonstrate, Huxley's films intertwine all three genres—instruction, propaganda, and even melodrama—suggesting the range of desires elicited, and satisfied, by the so-called scientific film.

Julian Huxley's Oscar-winning documentary, *The Private Life of the Gannet*, was filmed in the spring of 1934 at London Films. The rather coy and distinctly melodramatic title was chosen by film magnate Alexander Korda, who hoped to ride on the coattails of his recent crowd pleaser, *The Private Life of Henry VIII*. As Julian Huxley recalled this film in his memoirs, it was a remarkable crossover success, probably because it cloaked an almost prurient interest in sexuality under the guise of "scientific" research: "It had a long run in cinemas all over England and America, as well as proving useful to departments of Zoology by illustrating the breeding biology, the strange mutual displays of the birds, and their aerodynamic skill."[29]

The Private Life of the Gannet wittily used human courtship and reproductive rituals as the backdrop to its scientifically faithful recording of gannet mating behavior. It thus illustrated Nichols's first two principles—cinema as movement and the use of the mechanics of cinematography to satisfy desire. Both the title and the representational strategies of *The Private Life of the Gannet* invoked (and negated) the fundamental human desire to spy on the primal scene. Linking reproductive behavior (in birds) to reproductive behavior (in humans, even kings), it mediated (and distanced) both by the technical skill of "one of London Films's best cameramen, Borrodaile, [and] all the apparatus for filming" (210).

The mechanics of the camera are used in *The Private Life of the Gannet* to satisfy desire: in this case, the wish to see details of gannet mating behavior previously unavailable to human viewing. The film enacted Nichols's second principle by using a lavish variety of different camera angles and different kinds of shots to capture the range of gannet courtship and mating activities. Huxley recalls that the film opened with an aerial shot from an RAF plane of the "approach to the island— first a blur of white, gradually resolving itself into thousands of separate white dots—the birds on their nests," went on to show "closeups of the feeding of the young and even of a fledgling throwing itself off the grassy cliff into the sea," and concluded with a sequence "supplied by my old friend John Grierson, 'father' of documentary films, who chartered a herring-boat to take close-ups of a swarm of gannets diving for fish—a beautiful sequence in slow motion" (210–211).

The other film Julian Huxley made in the 1930s illustrated Nichols's third principle, the cinematic lie constructed by camera presentation and reception. *From Generation to Generation* was produced in 1937 by the Eugenics Society to educate its viewers in the principles of sound eugenic reproduction. Whereas the first film dealt with gannet reproduction, here it is human beings whose reproductive activity is the subject of scientific scrutiny. However, this film, screened for me in the Wellcome Institute for the History of Medicine where the Eugenics Society papers are held, has a "crossover" quality similar to that of *The Private Life of the Gannet.*" It evokes associations both to scientific documentaries and information videos (like the sex education films produced by Disney and inflicted on my generation of fifth-grade girls) while in its reproductive voyeurism it anticipates such recent hits as *Look Who's Talking*.[30]

The film attains its discussion of human reproduction, typically, via animal reproduction: an opening shot of rabbits is accompanied by a voice-over defining the sperm and the egg. Next follows the shot of a sea urchin and a discussion of heredity as expressed by coat color in cows and guinea pigs and by comb size and shape in fowls.[31] The discussion of animal heredity concludes with two familiar examples: the evolution from wolf to dog and the breeding of race horses, closing with a shot of two Derby winners: sire and son. The competitive theme continues as the film moves into the question of human heredity and the need for controlled breeding: "Eugenics seeks to apply the known laws of heredity so as to prevent its degeneration and improve its quality," explains the invisible narrator. A shot of the famous Phelps family, many of whom were champion boatmen on the Thames, plugs the eugenic message cheerily: the champions among the family members are marked in white.

The film becomes ominous, however, as it reaches its theme: how what it calls "mental deficiency" can be inherited. The facts as the film presents them are grim. We are shown a "normal" mother and father, who gave birth to seventeen children, five of whom died in infancy. The remaining seven—"mentally-deficient"—are shown first in a posed shot, and then as representations marked garishly in black on a mock-up of the family lineage. "Mental defectives are cared for in institutions all over this country," the film narrator intones, over a shot of women sitting and knitting. Shots of other people exercising outdoors are accompanied by the explanation that "mental-defectives" benefit from "physical training." The narrator provides a somber summary: although "high-grade mental-defectives" are normal looking and employable, they pose a danger to the community, for they may pass on their mental deficiency to their offspring. "Once born," the narrator concludes, "mental-defectives are happier and more useful in institutions . . . but better for them by far if they had never been born."[32]

From the portrait of hereditary "mental-deficiency," the film moves quickly back to happier ground, clearly desiring to leave its viewers admiring the values of positive eugenics rather than contemplating the difficulties of negative eugenics. The principle of inherited abilities, rather than inherited disabilities, is illustrated through the stories of two families: the Godfrey family, Royal Bandmasters in the Brigade of the Guards, and the illustrious Terry-Gielgud family, who as the narrator explains, have been "connected with arts and the stage for several generations."

We are then shown a family tree remarkable not only for its illustrious inclusions but also for its omissions. The film traces a line from Ellen Terry to her daughter Edith Craig, and then to Fred Terry and Julia Neilson, the film actress Hazel Neilson, and finally Frank Gielgud of the BBC and his son John Gielgud, the celebrated actor. Embodying, even concretizing, Nichols's third principle of the film medium—the camera lie—the film omits mention of both sexual deviancy (Ellen Terry's lesbianism) and illegitimacy (Terry's liaison with Gordon Craig, which produced Edith Craig). Rather than acknowledging variation even within a talented norm, the film opts for the illusion that perfect control—in this instance, of a family's intellectual inheritance and social performance—is possible, if only the family breeds eugenically. *From Generation to Generation* ends with the appearance of its authoritative, previously invisible narrator, who turns out to be Julian Huxley, standing in front of a chart mapping different levels of fertility. The film's survey of eugenics has only skirted the issue of negative eugenics—raising the question of restricting birth without addressing the mechanism for doing so—and it bids farewell from the firmer and more comfortable ground of positive eugenics. As Julian Huxley explains in conclusion, "If we are to maintain the race at a high level mentally and physically, everybody sound in body and mind should marry and have enough children to perpetuate their stock and carry on the race."[33]

To recapitulate: Julian Huxley's films adhere to the essential traits of the cinematic medium, as Robert Nichols explained them in the *Realist*. *From Generation to Generation* extends the interest in movement of Muybridge and Marey from an individual's physical movement in space to a trait's movement in time, as it passes from individual to individual over generations. Evolutionary movement is also represented, in the film's progression from animal to human heredity. In fact, C. P. Blacker asked Huxley to narrate *From Generation to Generation*, replacing Lord Horder, precisely because of his zoological expertise: "The first part of the film is about animals and the second part applies, as it were, to man the principles which can be observed operating in their simplest manner in animals. Your being a zoologist and anthropologist and also being Secretary to the Zoo carry guns which [Lord] Horder does not carry for the purpose of this particular battle."[34]

Huxley's eugenic film also relies on camera mechanics to express desire, no longer the wish to witness the primal scene (the emotion that undergirds *The Private Life of the Gannet*, if at one species remove), but now the desire for control that fueled both negative and positive eugenics. Finally, the film enacts the camera lie in a number of ways: it omits the irregularities in the Terry-Gielgud genealogy, and by presenting the "mental-defectives" as alien, it obscures an ongoing relationship between them and the Eugenics Society filmmakers.

That the "mental-defectives" pictured in the film were subject to control by the very act of filming is revealed by a curious anecdote. Apparently the Eugenics Society had a continuing relationship with the family who served in the film to illustrate mental-deficiency. Their appearance in the film was not candid, but contractual; as Tim Boon has pointed out, they were paid for their labors with the gift of a secondhand three-piece sofa set.[35] Indeed, the sway over them extended from the fiduciary to the psychological, for it seems they were persuaded by the very argument they used in the film to illustrate: that hereditary mental deficiency can be avoided by eugenic reproduction. One of the sons apparently approached Julian Huxley with the request that he serve as sperm donor to the son's wife, a request which, Huxley ruefully reported to Blacker, president of the Eugenics Society, he had decided not to honor.[36]

Julian Huxley's success with *The Private Life of the Gannet* inspired Aldous to look into selling film rights to his own work even before the filming of *From Generation to Generation*. A letter Aldous wrote to his brother in May 1935 illustrates the range of meanings that the cinema had for him in the years immediately after the publication of *Brave New World*. He wrote responding, it seems, to Julian's offer to act as midwife for a film deal: "If you see [Alexander] Korda, casually mention *Brave New World* by all means and see what his reactions are" (Smith, *Letters*, 395). But then the younger brother goes on to assert his independence of his older brother's help:

> I saw him several times this winter and he was very amiable. It is more
> or less arranged that Ted Kauffer and I shall do some shorts for him—
> on psychological themes such as Dreams (a bit of fun there, what with
> Freudian interpretations and trick photography); Handreading and
> graphology, and hands in general; survival of superstitions in modern
> times. . . . I wanted very much to do a short on Physique and Charac-
> ter and have indeed prepared it: but that requires a lot of people to act
> the different types of schizothymes and pyknics, or a great deal of field
> work; so for the moment that's being left aside. (395)

As the film shorts mentioned illustrate, in Aldous Huxley's mind cinematic principles could serve scientific, ethnographic, psychological, and medical investigations, recalling the early work of Muybridge, Marey, and Lumière. But with his mention of *Brave New World*, they are also linked to the twentieth century's most radical vision of reproductive technology.

EMBRYOS ARE LIKE PHOTOGRAPH FILM: *BRAVE NEW WORLD*

One of the most memorable passages in *Brave New World* (1932) is the nightmarish scene in the Central London Hatchery and Conditioning Centre's Embryo Store, where embryos gestate ectogenetically under dim, red light: "The sultry darkness . . . was visible and crimson, like the darkness of closed eyes on a summer's afternoon. The bulging flanks of row on receding row and tier above tier of bottles glinted with innumerable rubies, and among the rubies moved the dim red spectres of men and women with purple eyes and all the symptoms of lupus. The hum and rattle of machinery faintly stirred the air."[37] For readers today, that passage may evoke the vivid visual effects of films by David Cronenberg and Ridley Scott: the red-robed gynecologists in their lavish operating theater, the rain-slick, neon-spangled streets of postmodern Los Angeles.[38] It is no coincidence that this scene from *Brave New World* anticipates such films of reproduction gone wrong, producing cyborg replicants with artificial memories or human-fly chimeras, for like those films, Huxley's novel links VT to RT. The interchange with which the scene begins establishes a central trope for Huxley's novel: "'Embryos are like photograph film,' said Mr. Foster waggishly, as he pushed open the second door. 'They can only stand red light'" (6).

In comparing embryos to photograph film, Mr. Foster probably means that embryos, like photographs, need to develop, a process that for each is best carried on under dim light. But the representations of embryo culture in *Brave New World* extend the comparison, for they reveal a heavy dependence on a variety of visualization technologies, from microscopy to cinematography. Thus the tour of the Central London Hatchery and Conditioning Centre features "the yellow barrels of the microscopes" lit by the winter sun; the Director's description of how fertilized ova are transferred, in their embryonic culture solution, "onto the specially warmed slides of the microscopes," where they are "inspected for abnormalities"; and a description of how

X-rays are then used to trigger the process of embryo budding, or "bokanovskification," which produces up to ninety-six identical embryos (1–3).

Visualization technology figures prominently in Huxley's representation of embryo culture in the opening section of *Brave New World*. One way to explain that juxtaposition of RT and VT would be simply to say that it reflects the state of the art in embryological practice, which was heavily dependent on microscopy in the early twentieth century. Indeed, Huxley was highly accurate in his representation of embryology in *Brave New World*, according to the eminent embryologist Joseph Needham, who reviewed the novel in *Scrutiny*:

> *The biology is perfectly right*, and Mr. Huxley has included nothing in his book but what might be regarded as legitimate extrapolations from knowledge and power that we already have. Successful experiments are even now being made in the cultivation of small mammals in vitro, and one of the most horrible of Mr. Huxley's predictions, the production of numerous low-grade workers of precisely identical genetic constitution from one egg, is perfectly possible. Armadillos, parasitic insects, and even sea-urchins, if treated in the right way, do it now, and it is only a matter of time before it will be done with mammalian eggs. Many of us admit that as we walk along the street we dislike nine faces out of ten, but suppose that one of the nine were repeated sixty times.[39]

More than just faithful representation accounts for the prominence of the visual in these reproductive images, however. Needham's image for bokanovskification—the apparition of those (sixty) faces in the crowd—rings a frightening biological change on the urban street scene prominent in the poetry of Baudelaire, Pound, and Eliot; the novels of Woolf, Joyce, and Rhys; and the photography of Steichen, Atget, and Weegee. Revising a central modernist image, Needham asks us to consider what modern urban mass culture has to do with mass [re]production, the "standardization of the human product" pushed to "fantastic, though not perhaps impossible, extremes," as Huxley described it in his foreword to *Brave New World* (xiii).

Needham thinks of a city street when reviewing *Brave New World* because the urban street scene, like VT and RT, subjects the stroller to the experience of a loss of individuality. As William Chapman Sharpe put it, "The urban anonymity that makes individual city dwellers appear less individual and important, less part of the enormous community they inhabit, must inevitably diminish their auratic distance and uniqueness."[40] A classic essay on the modern urban street scene, Walter Benjamin's "Work of Art in the Age of Mechanical Reproduction" can help us understand the link that Needham makes, via the city street scene, between the new visualization technologies and the reproductive advances that Huxley portrays in *Brave New World*. Benjamin coins the term "aura" for that quality of uniqueness possessed by the individual which is obliterated by the process of standardized mass production, whether by photography, cinematography, industrial rationality, or Huxley's audacious new invention, Bokanovsky's Process. A loss of the aura results, according to Benjamin, when modern technologies like photography and cinematography intrude on the aesthetic domain previously dominated by painting.[41]

Benjamin expresses his sense of the radical contextual disjunction produced by the new technique of cinematography in a curious image drawn from medical science: a surgical operation (233). The difference between the painter and the cameraman is like that between a magician and a surgeon, Benjamin argues. As he explains, drawing on the work of French poet and essayist Luc Durtain, the difference lies in the extent to which the two "medical practitioners" preserve the aura of distance in their interactions with the patient's body.[42] The magician, cloaking himself in the power and authority of the medical practitioner, remains at a remove from the patient; in contrast, "the boldness of the cameraman is . . . comparable to that of the surgeon."[43] Benjamin portrays the surgeon's relation to the patient as paradoxically both socially distant and physiologically intimate: "The surgeon . . . at the decisive moment abstains from facing the patient man to man; rather, it is through the operation that he penetrates into him" (233).

Distance versus proximity; surface versus depth; superficial vision versus deep vision; emotional or social relations versus instrumental relations: the comparison between magician and surgeon or painter and cameraman that produces these differences results also in a "tremendous difference between the pictures they obtain," Benjamin observes (234). And this is where the mediation of visualization technology is crucial: with his camera the photographer challenges the auratic individuality of his subject, just as the surgeon with his surgical instruments challenges the body aura of his patient. Benjamin audaciously concludes that the cinematographic vision is preferable to the painterly vision for the same reason that the surgeon's deep encounter with the patient's body is an advance over the surface practices of the magician: both surgeon and cameraman use the mediation of instrumentation to obtain depth knowledge, and both assemble a "picture" of reality under a "new [and preferable] law."[44] The aesthetic impact of cinematography (the "new law" of the cameraman) is comparable, Benjamin's analogy implies, to the biomedical impact of the seventeenth-century scientific revolution, in which magic was vanquished by the new law of scientific empiricism.

Giving fleshly form to Benjamin's idea of the loss of the aura, *Brave New World* expresses Huxley's mandarin anxieties that a debased humanity would follow from new technologies of mechanical reproduction.[45] The novel's beginning image of the Embryo Store embodies that anxiety concretely, as microscopists deliberately engineer the embryos to restrict their potential. The novel also expresses those anxieties powerfully in its ending: but this time the emphasis falls not on reproductive, but on visualization technology, and its goal is less the [re]production of sameness than the policing of difference: constructed as the allegiance to a culture of magic.[46]

David King Dunaway has argued that John the Savage's suicide at the end of *Brave New World* reworks the traumatic death of Aldous Huxley's younger brother Trevenen. Forbidden to marry the family parlor maid with whom he had fallen in love, Trevenen hanged himself at the age of twenty-four.[47] Both suicides can be understood as a form of reproductive self-policing: Trevenen Huxley cutting short his romance with the parlor maid who had received no more than a village schooling, and John the Savage resisting the allure of "Lenina whom he had promised to forget" (171). But the conclusion of *Brave New World* invokes Aldous Huxley's older brother's interests as well. Just as Julian Huxley used cinematography to oversee animal and human reproduction in *The Private Life of the Gannet* and *From Generation to Generation*, so Aldous represents cinematography functioning first as voyeur, and then as trigger, to John's ritualistic final acts: flogging himself, and finally killing Lenina, with the rawhide whip.

John's self-inflicted flogging is caught "from his carefully constructed hide in the wood three hundred metres away, [by] Darwin Bonaparte, the Feely Corporation's most expert big game photographer" (172). This episode of cinematic voyeurism links ethnographic and zoological uses of film to its sociological and political functions: the construction of an endless mass market for titillation, and the legitimation of state power by the delegitimation of other cultural practices, ethnicities, nations, even species.

The very name of the cameraman, Darwin Bonaparte, indicates his role as embodiment of evolutionary and nationalistic ascendancy. As

Fatimah Tobing Rony has observed in her study of ethnographic cinema, "The desire to demarcate difference and the quest to describe the body according to racial types coincided with the rise of imperialism and nationalism: the discourses of race, nation, and imperialism were intimately linked" ("Iconography of Race," 266). Film was the preeminent tool for demarcating difference and classifying racial types at the turn of the century, as Rony has pointed out: "Cinema was, and is, intimately linked with the evolutionary ideology that posits race as the predominant means of explaining human difference" (264). Darwin Bonaparte embodies this ideological investment in an eugenic and evolutionary cinematography. So, as he films John whipping himself, Darwin Bonaparte exults that this is his greatest moment "since his taking of the famous all-howling stereoscopic feely of the gorillas' wedding" (172). Moreover, in an uncanny anticipation of the film Julian Huxley would make two years later, *The Private Life of the Gannet*, Darwin Bonaparte produces his magnum opus, a film "almost as good . . . as the *Sperm Whale's Love-Life*—and that, by Ford, was saying a good deal!" (172–173).

The passage in which Bonaparte films what would become the hit feely, *The Savage of Surrey*, comprises a painfully sardonic send-up of the controlling scientific gaze and clinical distance of cinematographers:

> He kept his telescopic cameras carefully aimed—glued to their moving objective; clapped on a higher power to get a close-up of the frantic and distorted face (admirable!); switched over, for half a minute, to slow motion (an exquisitely comical effect, he promised himself); listened in, meanwhile, to the blows, the groans, the wild and raving words that were being recorded on the sound-track at the edge of his film, tried the effect of a little amplification (yes, that was decidedly better); was delighted to hear, in a momentary lull, the shrill singing of a lark; wished the Savage would turn round so that he could get a good close-up of the blood on his back—and almost instantly (what astonishing luck!) the accommodating fellow did turn round, and he was able to take a perfect close-up. (172)

This filming of John's self-flagellation ironically recapitulates the earlier scene in which Bernard and Lenina, as tourists, witness the snake dance in the pueblo of Malpais. There is a striking symmetry between the two scenes, illustrating how ethnographic cinema, like tourism, produces a sense of cultural primacy by a positioning and viewing of the ethnic Other.[48] We can apply to *The Savage of Surrey* Rony's comments on Regnault's ethnographic films in the 1890s: "The visitor [viewer] was in a sense invited to act as a scientist and colonialist, to acquire knowledge by looking at the body and its habitat. He was also invited to engage in sexual voyeurism: the exhibition [film screening] was the site for the viewing of the *unseen*" ("Iconography of Race," 272).

John has been tracked, filmed, and turned into a commodity no less than the Malpais snake dancer, who serves as an attraction for the British tourist industry, or the sperm whale, who fuels first the whaling industry (lighting the lamps of England), and then the film industry. As the feely in which he figures is consumed by a public voracious for self-confirming images ("seen, heard and felt in every first-class feely-palace in Western Europe"), his complex individual being is reduced to the representative status of an interchangeable primitive: he becomes "the Savage of Surrey" (173). Bonaparte's feely illustrates Rony's observation that "in ethnographic film generally, the viewer is confronted with images of people who are not meant to be seen as individuals, but as specimens of race and culture, specimens that provide the viewer with a visualization of the evolutionary past" (265). Yet by his cross-cutting between the camera image and the cinematographer's aesthetic manipulations and evaluations, Huxley raises the crucial question: Who really is the Savage? And in the concluding image of John's dangling feet, "like two unhurried compass needles," he problematizes the division of the world into "the old country" and the "brave new world."

The beginning and end of *Brave New World* converge in their portrait of two different techniques of mechanical reproduction, reproductive technology and visualization technology, both of which produce powerful challenges to the aura of the individual and the work of art. The Central London Hatchery and Conditioning Centre with which the novel opens manufactures so-called highly evolved, "civilized" man by a surgical and psychological intrusion into the very body and environment of the embryo. And Darwin Bonaparte's feely with which the novel ends produces so-called primitive man, the "Savage of Surrey," by a parallel cinematographic intrusion into John's passion and his pain.

BRAVE NEW WORLD: THE UNMADE MOVIE

Ten years after he had given his brother the go-ahead to sound out Alexander Korda about a film deal for *Brave New World*, Aldous Huxley moved to Hollywood. As Dunaway tells us, by 1945 he was riding high on the successes of his scripts for *Madame Curie*, *Pride and Prejudice*, and *Jane Eyre* (*Huxley in Hollywood*, 202–203). Fired with the desire to turn *Brave New World* into a movie, Huxley sketched out the screen-play in a letter to his close friend, the screenwriter Anita Loos.[49] This letter offers a valuable glimpse of how Huxley juggled his ongoing fascination with both RT and VT. Revealing his negotiations with the new medium of film, it also gives us a sense of his retrospective understanding of the central meaning of *Brave New World* and previews the concerns that would shape his novel to come, *Ape and Essence*.

Huxley emphasizes two points in his plan for the script of *Brave New World*: the requirements of the film medium, and his desire to remain faithful to "the point of the original story . . . namely, that the really revolutionary changes will come about from advances in biology and psychology, not from advances in physics."[50] Huxley explains to Loos the necessity "for film purposes, to write the scenes of the future in the form of cut-forwards from a contemporary starting point" and to provide the audience with sufficient orientation and explanation "to show that even the most extravagant pieces of satiric phantasy stem inevitably and logically from present-day seeds and are the natural end-product of present-day tendencies."[51] Huxley acknowledges that "something about the atom bomb will of course have to be brought in" (534). But he goes on to enumerate for Loos some of the biologically grounded possibilities for the future that he has simply taken from "the practices of contemporary dictatorships": "artificial insemination of women with the semen of racially pure sires, selective breeding . . ." (535).

Mapping out for Loos a script that would move from a contemporary starting point into the future, Huxley generates a plot that strikingly resembles the satiric portrait of J.B.S. Haldane as Shearwater in *Antic Hay*, while incorporating his brother Julian's scientific interests—experimental zoology and embryology:[52]

> I had thought vaguely of making it revolve around the person of a very clever but physically unattractive scientist, desperately trying to make a gorgeous blonde, who is repelled by his pimples but fascinated by the intelligence of his conversation, as he steers her round his laboratory, showing her artificially impregnated rabbits, newts with tails grafted on to the stumps of legs. . . . He holds forth about the possibilities of applied biology—babies in bottles etc—and we cut forward to the story, from which, at intervals, we cut back to his contemporary comments, which serve as a kind of Greek Chorus to the drama of the future.[53]

Like Shearwater, who as Haldane himself acknowledged "is endowed with a hopeless passion for a not really inaccessible lady," the physically unattractive scientist in the projected screenplay Huxley described to Anita Loos is rejected—"left disconsolate among the white mice and the rabbit ova" (535). But Huxley modifies that conclusion in a troubling afterthought that anticipates his portrait of Einstein in *Ape and Essence*: the "physically unattractive scientist" is "an emblem of personal frustration who is yet the most revolutionary and subversive force in the modern world" (535).[54]

Huxley never made this movie of *Brave New World*, although he discussed plans to do so at length with Burgess Meredith (who wanted to act in it) and Paulette Goddard, who wanted to finance it. The funding was tangled, according to Dunaway: Ralph Pinker, the "British agent who had gone bankrupt on Huxley, had sold the film rights [to *Brave New World*] to RKO for the pitiful sum of £750. RKO wanted $50,000 to resell the rights, and the picture had to be dropped" (207).[55] But the letter to Anita Loos suggests another reason why *Brave New World* came to nothing: the increasing power of censors to control and police the film image. "One practical point worries me," Huxley confided to Loos, "what will the Hays Office say about babies in bottles? We must have them, since no other symbol of the triumph of science over nature is anything like as effective as this. But will they allow it?"[56]

Huxley's attempt to imagine the filming of *Brave New World* brought back an insight he expressed two years earlier, in a work of nonfiction: the realization that power resides in, and constitutes itself through the manipulation of, vision. The Hays Office censors could refuse to allow the image of "babies in bottles" in the script, and in so doing could protect the institution of science against the film's critique of its power. Two years earlier, in *The Art of Seeing* (1943), Huxley challenged another sort of scientific hegemony: doctors' control over the treatment of vision problems. There, he argued passionately that doctors have wrongly arrogated to themselves the power to treat failing vision.[57] People can learn to control and improve their own visual powers, Huxley argues; they are only harmed by the institutional hegemony of organized medical science, which has produced power for itself by restrictively defining the issues affecting eyesight. "Ever since ophthalmology became a science, its practitioners have been obsessively preoccupied with only one aspect of the total, complex process of seeing—the physiological" (vii). Huxley's analysis in *The Art of Seeing* and his response to the power of the Hays Office censors, joined with his appreciation in *Brave New World* of how reproductive control epitomized scientific mastery over nature, led him to an understanding of bio-power: an appreciation of how the control of bodily processes (whether of visualization or reproduction) can produce power. Four years later, he would give that analysis fictional form in *Ape and Essence* (1949).

Brave New World focuses on RT against the background of VT, but *Ape and Essence* uses the new visualization technologies of the cinema and astronomy to illuminate a reproductive plot. No babies in bottles in this novel, however: the trajectory of *Ape and Essence* is, as its title indicates, regressive rather than progressive.[58] Instead, we have a "manuscript-found-in-the-bottle" and what might be called a repro- ductive reduction: humanity reduced to the level of the ape, no longer making love by choice, but engaging in a seasonal rut.[59] The plot is double-layered: at its core, the love story of Alfred Poole the botanist and Loola, a young woman who has somehow escaped the regression to animal rut induced in her fellow countrymen by the genetic defor- mations caused by atomic and biochemical war. Enveloping that anti- romantic romance is the story of its resurrection, as the abandoned film script is found by the unnamed narrator and his friend, the Holly- wood screenwriter Bob Briggs. The twinned influences of an industry and a person, of self-absorbed Hollywood cinema and the astronomer Edwin Hubble, are central to Huxley's investigation of "the power of science over nature" in *Ape and Essence*.

Hollywood plays a central role in Huxley's hybrid novel, both in the core plot dramatized by the (rejected) screenplay of William Tallis as well as in the explanatory autobiographical narrative. In its two parts, the novel bridges early- and late-twentieth-century uses of the cinema, as well as embodying Nichols's three cinematic principles of move- ment, satisfaction of desire, and the camera lie. The script alludes to early-twentieth-century ethnographic and zoological constructions of the racial and species other, while reversing them, because the ex- plorers come from New Zealand, a land "like Equatorial Africa . . . too remote to be worth anybody's while to obliterate" (31–32). *Ape and Es- sence* thus extrapolates from the conclusion of *Brave New World*: civili- zation has become savagery, and so-called savage customs now reign in hallowed halls. "What splendid tribal dances in the bat-infested halls of the Mother of Parliaments! And the labyrinth of the Vatican—what a capital place in which to celebrate the lingering and complex rites of female circumcision!" (32). Recording the New Zealand Re-Discovery Expedition to North America, a land made barbarous by the effects of atomic warfare and pollution, the script follows Chief Botanist Dr. Poole, as he encounters the mutated flora and fauna of this post-nuclear continent, including the barbaric human sociosexual practices of the ape-like Americans.[60]

The autobiographical introduction invokes mid-twentieth-century practices of film censorship to police moral and political boundaries. Huxley recounts the fate of Tallis's rejected, discarded, and rediscovered script against the implicit background of the 1947 Waldorf Statement. Issued in response to the House Un-American Activities Committee hearings into Communist infiltration of the film industry, this communiqué by the Association of Motion Picture Producers attempted to ward off external regulation of the film industry by inviting "the Hollywood talent guilds to work with us to eliminate any subversives; to protect the innocent; and to safeguard free speech and a free screen wherever threatened."[61]

The Waldorf Statement's Orwellian advocacy of voluntary intellectual enslavement in defense of freedom is echoed by the novel's central aesthetic principle—the cinematic lie: what you see is (not). This lie can be seen in the script-within-a-novel, which contains its own cancellation, much as does the Waldorf Statement. Tallis, the author, impossibly concludes his work with the scene of his own death. Yet it is precisely the mysterious circumstances of Tallis's death that separates his script from the pile of scripts destined for the incinerator (21). The cinematic lie is further embodied in the script's insistent, self-conscious representation of visual control, recalling Nichols's third principle: "In what we present to the camera and how we make the camera receive it lies the art of the motion picture. . . . For the peculiarity and beauty of the camera is that it can lie and keep on lying" ("Movies as Medium," 148–49). Control of the visual is foundational to Tallis's script, which combines voice-over narrative with obtrusive camera directions throughout: "the Camera moves across the sky," "telescopic shot of low hills," "dissolve to a series of montage shots" (*Ape and Essence*, 31, 38, 142).

The specific points of view indicated in many of the scenes are usually mediated through some form of visualization apparatus: the binoculars with which the New Zealanders first see North America; the magnifying glass with which Dr. Poole, chief botanist, encounters its mutated flora; the Navy night-vision binoculars with which Poole and the Arch-Vicar watch the orgy of purification; the television news reels with which (the narrator tells us) we will watch "the skinning alive, in full colour, of the seventy thousand persons suspected at Tegucigalpa, of un-Honduranean activities," the "stainless steel barrel of the microscope" in Dr. Poole's abandoned laboratory, or merely Dr. Poole's own eyes, as he watches Loola walking away from him: "Trucking shot from Dr. Poole's viewpoint. NO, NO. NO NO . . ."

Whether the focus is minute or distant, the purpose of all of these looks is the exercise of control. A passage midway through the novel, in which the Arch-Vicar explains the triumph of "Man . . . against Nature, the Ego against the Order of Things," explains how control through visualization operates at all levels, from the microcosm to the macrocosm, in order to produce the scientific control of nature (87).

> Dissolve to a shot through a powerful microscope of spermatozoa frantically struggling to reach their Final End, the vast moon-like ovum in the top left-hand corner of the slide. . . . Cut to an aerial view of London in 1800. Then back to the Darwinian race for survival and self-perpetuation. Then to a view of London in 1900—and again to the spermatozoa—and again to London, as the German airmen saw it in 1940. (89)

As this passage indicates, central to *Ape and Essence* is an analogy between self, nation, and species engaged in the drive for self-perpetuation, or to put it another way, a connection between human population growth and warfare. Huxley imagines a future in which atomic war has changed human methods of reproduction. However, the result is far from the rationally controlled production of scientifically engineered embryos, as in *Brave New World* or (some argue) the Human Fertilisation and Embryology Bill nearly half a century later. Rather, as he confided his plan for the novel to Anita Loos: "I may write something about the future . . . about, among other things, a post-atomic-war society in which the chief effect of the gamma radiations had been to produce a race of men and women who don't make love all the year round but have a brief mating season. The effect of this on politics, religion, ethics etc would be something very interesting and amusing to work out."[62]

The passage from *Ape and Essence* indicates the far-reaching implications of this shift in reproductive practices. Interweaving microscopic interventions into conception with generalized programs of population control, evolutionary struggle with nationalistic war, an image of the sperm reaching the egg with a moon shot, the passage illustrates what Zoe Sophia has called the "sexo-semiotics of extraterrestrialism": "the condensation of extinction anxieties onto ambiguous symbols, and their displacement onto other political and moral issues," in particular reproduction. "Within the sexo-semiotics of technology," Sophia observes, "every tool has reproductive implications and represents a form of reproductive choice: *every technology is a reproductive technology*."[63]

In its metaphoric comparison of the microscopic journey of sperm to egg with the telescopic trip to the moon, this passage also indicates the novel's debt to Huxley's close friend, the astronomer Edwin Hubble. Hubble was, in Huxley's view, "the man who, more than any other single individual, is responsible for [the] new revolution in cosmological thought. . . . The universe as we know it today—infinitely improbable, fathomlessly mysterious—rests in the main on Hubble's observations."[64] Hubble discovered the existence of galaxies other than our own, beginning with the Andromeda nebula, using the massive Mount Wilson and Palomar telescopes.[65] He formulated Hubble's Law—"the more distant a galaxy was, the greater was its speed of recession"—leading to the understanding that the universe itself was expanding.[66] His dramatic astronomical findings "extended the horizons of the universe to 500 million light-years," thus producing a rearrangement of the human position in the universe greater even than the Copernican revolution.[67]

Huxley uses Hubble—and the astronomy he represents—to critique Hollywood in *Ape and Essence*. The novel begins with a veiled allusion to Hubble's revolutionary discovery that the universe was both older and far vaster than had previously been understood. That discovery has been compared to Copernicus's revolutionary reconceptualization of astronomy. Yet despite their humbling consequences for humanity, the novel's opening sequence ironically asserts, neither Hubble's discovery nor the Copernican revolution that proceeded it can shake the insistent egotism of Hollywood: "In spite of all the astronomers can say, Ptolemy was perfectly right: the center of the universe is here [in Hollywood], not there" (7). To narcissistic Hollywood, and its denizens screenwriter Bob Briggs and producer Lou Lublin, the Ptolemaic system *is* still in force: they are the centers of their tiny universes. The novel's opening establishes its concern with two conflicting constructions of the human being: Hubble's new cosmology and Hollywood's cinematography.

As the screenplay-within-a-novel unfolds, Huxley implicitly asks the reader to borrow Hubble's telescope as well as the binoculars of the New Zealand Re-Discovery Expedition: to see the universe afresh, and rethink the human position within it. In his unpublished essay, "Stars and the Man," Huxley addressed that issue, considering Hubble's impact upon our understanding of the universe. The passage not only recapitulates the central themes of *Ape and Essence*, but it uncannily anticipates Sir Ian Lloyd's speech to the House of Commons over thirty years later, with which this chapter began:

> The effects of any given scientific discovery upon the philosophy by which men actually live are hard to predict. How precisely will man's world view be modified by this latest, this more than Copernican revolution, for which the great telescopes and their users, above all Edwin Hubble, have been responsible? Bulldozers and atom bombs have intoxicated us with the illusion that man is virtually omnipotent. On the other hand, cybernetics and biochemistry have inclined us to the belief that individual human beings are merely the by-products of molecular arrangements and subatomic events. Will the new cosmology confirm us in that sense of personal insignificance, which is the paradoxical corollary of our collective bumptiousness?[68]

As Huxley's response to Hubble suggests, *Ape and Essence* explores the challenge to human identity posed by the confluence of a range of scientific advances, from the great telescopes and the atom bomb to cybernetics and biochemistry. Common to all of these advances is the technique described by Hermann Muller: control of the macrocosm through control of the microcosm. Like Sir Ian Lloyd after him, Huxley sees these new technologies as linked by their common challenge to human identity, torn as it is between good and evil, a sense of "personal insignificance," and the sense of group importance Huxley calls "collective bumptiousness." But to understand fully why Lloyd, like Huxley in *Ape and Essence*, links such visualization technologies as the telescope and the microscope to reproductive technologies like embryo experimentation, we must turn to one final text: the scientific memoir of Doctors Robert Edwards and Patrick Steptoe. *A Matter of Life* (1981) recounts their discovery of the technique for in vitro fertilization, or the creation of test-tube babies.[69]

RT AND VT IN THE RACE FOR IVF

We have seen how Aldous Huxley's fiction reflected the scientific interests of his brother Julian, from his zoological work in the twenties to his ornithological and eugenic observations, mediated by film, in the 1930s. As Aldous observed in *Literature and Science* (1963), however, the influence was always bidirectional. Robert Edwards and Patrick Steptoe's narrative of the development of in vitro fertilization, the "test-tube baby" technology that concretized Huxley's imaginings in

Brave New World, as I argued in Chapter 1, reveals scientific practices saturated by the reproductive metaphors and the passionate concern with powers of visualization that characterized the writings of Aldous Huxley. Of course, to invoke a division between the two cultures is to participate in an obsolete construction of the relationship of literature and science. We have learned from Latour and Woolgar, among others, the powerful role in scientific practice played by narrative, metaphor, analogy, and myth, and Hayles, Beer, and Levine, among others, have sensitized us to the scientificity of much fiction.[70] Edwards and Steptoe's narrative is itself a member of a specific literary genre and thus shaped by generic conventions and intertextualities, whereas Huxley's interest in visualization tested the boundaries of medical science when it resulted in his book of alternative therapy for ocular disease, *The Art of Seeing.* Yet if we follow out Huxley's literary themes and concerns in the scientific narrative of Edwards and Steptoe, we can see that a deep cultural connection between reproductive technology and visualization technology preexisted, and left its traces on, the development of in vitro fertilization.

Embodying something like a textual double-helix, *A Matter of Life* interweaves two narrative strands: Edwards's scientific education in embryology leading to an interest in fertilization outside the uterus, and Steptoe's independent development and refinement of techniques that would enable the two of them to collaborate on the process eventually leading to the successful creation of Louise Brown, the world's first "test-tube," or in vitro baby. A quick tracing of each narrative strand will reveal some remarkable recapitulations, in this joint scientific narrative, of issues central to Huxley's literary works discussed earlier.

Casting his journey to the development of IVF as a bildungsroman, Edwards defines himself as having "long been interested in the scientific processes of reproduction" (7). As if echoing Huxley's image in *Ape and Essence* comparing sperm under the microscope to London under the blitz, Edwards traces that passionate interest in reproduction back to his childhood "wartime days [when] I asked myself schoolboy questions about fertilisation and birth" (7). The next stage in that developmental journey occurs in "the Department of Zoology [when] I asked more complicated questions as I attended tutorials or lectures on fertilisation and the early stages of animal life" (7).

> On one occasion in the Zoology lab I looked up from a microscope thinking. *Why does only one spermatozoon enter an egg?*

"A million million spermatozoa," wrote Aldous Huxley, "all of them alive," and he continued amusingly:

> Out of this cataclysm but one poor Noah
> Dare hope to survive.
> And among that billion minus one
> Might have chanced to be
> Shakespeare, another Newton, a new Donne—
> But that One was me.

When the summer examinations arrived, the One that was Me did not do well. (7–8)

With the bad examination results of "the One that was Me," Edwards not only places himself at a developmental crossroads, but he links himself, through a chain of identifications beginning with the "froward Homunculus," to his imagined creator, Aldous Huxley. As Edwards constructs this turning point, it leads to his decision to abandon his previous field of agriculture, and to apply to study in the embryological laboratory of the great C. H. Waddington, in Edinburgh. "I was the clever, ambitious, scholarship boy who looked as if he had now fallen flat on his face. I expected no reprieve. Imagine my pleasure then, my sense of relief, when I was accepted at the Institute" (9). Waddington's Institute of Animal Genetics becomes the site of Edwards's next intellectual turning point, a lecture by Alan Beatty,

> an expert on fertilisation and the development of embryos in mice. He had recently made a film called "Inovulation" with the Film Unit of the Institute. In it he showed how a fertilised mouse egg had been taken from the uterus (womb) of one mouse and injected into the uterus of another mouse. The fertilised egg had subsequently grown into a foetus and then into a healthy newborn mouse. (11)

This film of mouse surrogacy—as we could now describe it— provided a galvanizing image for the young Edwards, one that determined the direction of his dissertation research. He decided to "follow the tenuous, exciting leads of my predecessors who sometimes had worked years before, decades before, even centuries earlier. They too had been interested in fathoming the secrets of fertilisation and in delving into the mysteries of the newly formed and developing embryo" (12). The links between visualization technology and reproductive technology are accumulating in Edwards's narrative: from the early link between microscopy and interest in the sperm, to this later role of cinematography in catalyzing Edwards's interest in attempting in vitro creation of an embryo. We can even find an echo of Huxley's metaphor of the creation narrative in "Fifth Philosopher's Song," when Edwards recalls of his years in Edinburgh, "I felt eventually that the numinous and the mysterious could be found . . . in the laboratory where each night I peered through a microscope at primitive sex cells" (16).

The VT/RT linkage returns, along with Edwards's identification with Aldous Huxley, at the first "breakthrough moment" in Edwards's personal developmental saga. In 1965, Edwards recounts, he worked late night after night, often to the sound of "rock music coming from the Regal Cinema," trying to fertilize human eggs with his own sperm. Finally, to his delight, he succeeded: "One spermatozoon had passed through the outer membrane of an egg" (50). A trip to the United States intervened, where working with Victor McKusick of Johns Hopkins Hospital he attempted, unsuccessfully, to replicate that achievement. Although no fertilization occurred during his work in the United States with ripening human ova, when he published his data in the *Lancet* the popular press made the inferential leap: "BIRTHS MAY BE BY PROXY. Underneath it my own name featured in a story about 'experiments reminiscent of Aldous Huxley's *Brave New World*'" (55).

The culmination of Edwards's solitary investigations—the meeting with Patrick Steptoe that would finally lead to their successful research program—occurs, once again, in a context that links reproductive technology to visualization technology. In a chapter invoking all the clichés of heroic scientific discovery narratives, and even titled "Eureka," Edwards recounts how, attending a convention of gynecologists in London, his attention is caught by an exchange concerning the clarity and effectiveness of laparoscopy as a technique for seeing inside the uterus. In a meeting featuring "a small cinema-like screen," the participants agree that it would greatly aid medical practice if "ovaries could be inspected easily. . . . We'd see how many eggs were growing. Perhaps the new method of laparoscopy could be of use here?" (60). One gynecologist dogmatically asserts, "Laparoscopy is of no use whatsoever. It is impossible to visualise the ovary using that technique" (61). Suddenly, dramatically, he is answered by another speaker from the back of the hall:

Forcefully, he recounted how, through laparoscopy, not only the ovaries could be seen, but also the Fallopian tubes and other parts of the reproductive tract. "Indeed," he continued, "the whole abdominal cavity can be inspected. I carry out laparoscopy routinely every day—many times over." . . . This obviously was *the* Patrick Steptoe of Oldham General Hospital. I felt immediately that here was a man I could trust and respect and work with. . . . He was utterly convincing and he offered to demonstrate, on the screen, the slides he had brought along to substantiate his claims. (61)

As Edwards's narrative twines around Steptoe's, the role of visualization technology becomes even more central. Steptoe also tells his story as a medical bildungsroman, but whereas Edwards's uses the Huxleyan interest in spermatozoa and fertilization as his point of origin, Steptoe traces a line from his boyhood in Witney, helping "the resident pianist at the local Palace cinema by playing, during matinees, the incidental music for the silent films of Tom Mix, Harold Lloyd, Raymon Novarro, Rudolph Valentino," to the moment of triumph when he flashed his corroborative color slides onto the screen. "I was able to take better shots inside the abdomen than I could outside with an ordinary camera," Steptoe recalls (65, 78). Not surprisingly, his visual display is spellbinding: as he narrates that moment, "There was another hush during the showing of these slides, and afterwards warm applause" (79). "Someone—I believe it was Elliot Philipp, a senior gynecological colleague from the Royal Northern Hospital, London—stood up and said, 'We must congratulate Mr Steptoe on those beautiful photographs'" (79).

The pleasures of looking (particularly at women's bodies) are incorporated in Steptoe's narrative of medical education from the very beginning. The moment that he constructs as responsible for his decision to become a gynecologist is the case of a "lovely red-haired girl [who] . . . had been the victim of a septic criminal abortion . . . [and] developed tetanus, from which she died" (64). "At her post mortem," Steptoe recounts, "I felt angry before her body on the slab—her beautiful figure, the pale face at last at rest, the striking red hair—angry as I observed the pathologist demonstrate how her abortion had been cruelly botched" (64–65).

Scholars have analyzed the semiotic excess that typically characterizes anatomical drawings and models that figure women: the erotic charge conveyed by the corpse's pose and the detailed presentation of the hair and skin; the way such purportedly objective representations have functioned to [re]enforce gender relations in society at large.[71] Steptoe's image of the dramatic case that made him a gynecologist participates in a representational tradition that links the "representation of a woman's body in the process of being dissected" with "the idea of unveiling," revealing the gendered nature of modern medicine. As Ludmilla Jordanova observes, "Woman, as the personification of nature, was the appropriate corpse for anatomy, which was not just literally male in that its exponents were men, but was symbolically male in that science was also the masculine practice of looking, analyzing, and interpreting."[72]

Steptoe exploits and maintains this linkage between medicine and the invasive look into a woman's body as he narrates his attempt to find a way around laparotomy—"an operation in which the abdominal cavity of the patient is opened up primarily for diagnostic reasons—so that the surgeon can directly visualize and feel the tissues and organs laid bare" (67). He tells of his encounter in 1925 with the work of Rendle Short, who had used a cystoscope to peer inside the bladder; his interest in Albert Decker's use of an instrument he called a telescope, in 1940, to look inside the female pelvis, approaching it through the vagina; and finally of his exciting discovery of Raoul Palmer's innovative use of a laparoscope, an "instrument [that] worked like a telescope, with an eye-piece to look through and with a lens inside the abdomen which permitted a view of the body organs" (69–70).

A trip to Palmer's hospital in Paris to view Palmer's "smash-and-grab surgical procedure" revealed that "laparoscopy gave a magnificent view inside the female pelvis," but Steptoe's enthusiasm for Palmer's work was succeeded by a "film on laparoscopy by Hans Frangenheim," which convinced him that "Frangenheim had the advantage of being backed by the lens-manufacturing companies of Germany so his laparoscope was more refined than the one used in Paris" (70–71). This nationalistically tinged quest finally ended when—like a scopophilic Frankenstein—Steptoe returned to Oldham "with the new German equipment they allowed me to buy, [and] was practising my technique on fresh cadavers awaiting post mortem in the mortuary. . . . Beforehand, of course, I had to ask permission to practice laparoscopy on the dead. . . . What a business that was!" (71).

If the culmination of Edwards's solitary journey was his successful fertilization of an ova in vitro with his own sperm, the culmination of Steptoe's journey was his successful look into the uterus of a living woman, with his own laparoscope. His narrative constructs it as a victory over deep discouragement, and situates that victory at the piano, as if in veiled allusion to his early role as accompanist to a very different sort of silent movie:

When in certain moods I would frequently find myself at the piano.
That evening, at home, was no exception. For two years I had hoped to
introduce into our unit a method to improve our diagnostic
methods. . . . I had studied culdoscopy and laparoscopy. . . . I had
practiced diligently on those dead ladies in the mortuary. And now I
had come to this blind end. I put down the piano lid and thought of
our old family saying that I had heard my mother repeat so often:
"Obstacles which you meet are really opportunities in disguise."

John Hirst persuaded the nurse to allow me to repeat the laparoscopy. "So it's up to you, now, Patrick," he said.

That second attempt was most successful. I was able to see clearly
the size, texture and colour of the whole uterus. (72)

That breakthrough moment, from "blind end" to unimpeded vision, when Steptoe's eye is able to penetrate without impediment into
the female interior, is followed by characteristic steps to consolidate his
technological accomplishment. The conclusion to his solitary narrative, "Laparoscopy Comes of Age," contains classic elements delineated by scholars of scientific practice. Steptoe incorporates *refinements
in his technique* ("a thin rod of quartz [was] incorporated in the laparoscope which was able to carry light cooled by an electric fan . . . to the
tip of the instrument within the abdomen," thus making it possible to
take color photographs for permanent records); he engages in *local
and nationalistic rivalries* over his scientific procedures; he writes a *textbook* on laparoscopy, hoping to convince his British colleagues of the
technique's surgical value (73, 77). And it is then, having surmounted
his own personal quest for deep vision, that Steptoe encounters Edwards, who has surmounted *his* own personal question for external
fertilization. The romance of IVF is now ready to unfold. The narrative casts the gynecological meeting as a dramatic encounter: across a
crowded meeting room, Edwards's skill in fertilization comes together
with Steptoe's skill at seeing within women's bodies.

This meeting of Edwards and Steptoe recalls Charlotte Haldane's
narrative of her meeting with J.B.S. Haldane, for both are written in
the romance genre, and both have at their center a striking image of
reproductive innovation. Whereas Charlotte Haldane's union with
J.B.S. was catalyzed by the image of ectogenesis, the Edwards-Steptoe
partnership was catalyzed by the technical processes of laparoscopy
and egg and embryo culture. We could even take this parallel a bit far-
ther, for both relationships were held together by what Charlotte
called the "philoprogenitive desire" (theirs or, in the case of the doctors, theirs intellectually, and their patients' emotionally and biolog-

ically). Although that desire to have children was never satisfied in the Haldane romance, in the Edwards-Steptoe partnership it is an inescapable fact that their successful accomplishment of IVF satisfied the unconscious fantasy of a joint male pregnancy.[73]

From that crucial meeting over the laparoscopic slides, the rest of the narrative of IVF has a familiar scientific up-curve; *A Matter of Life* now tells of the slow surmounting of obstacles, while obtaining the crucial photographic documentation for every accomplishment.[74] Whereas the rest of Edwards and Steptoe's narrative has a metaphoric depth and richness that merits equally close examination, I will conclude this discussion with a moment that crystallizes the RT/VT connection this chapter has explored. The chapter "Four Beautiful Human Blastocysts" recounts Edwards and Steptoe's success in bringing four human embryos to blastocyst stage—the "last stage of growth before the embryo begins its implantation in the mother's womb." As Edwards recounts it, he stared into the microscope at "an unbelievable sight: four beautiful human blastocysts . . . the beginning of the foetus as it started its journey towards life. . . . I knew that instant that we had reached our goal: the early stages of human life were all there in our culture fluids, just as we wanted" (100–101). The image with which that crucial chapter ends participates in the same in-and-out-and-up visual movement, that same linkage of microscopy, reproductive technology, and telescopy that we have seen in Sir Ian Lloyd's metaphor and Aldous Huxley's writings, including those on Edwin Hubble. Edwards walks out of the laboratory into the night, on his way to his mother's house: "I looked up at all the stars, the moon, the night sky over Oldham, and considered the equally amazing sights I had just seen under my microscope" (102).

THE FETUS/WOMAN DISJUNCTION: DESIRE AND THE LIE

The narrative of Edwards and Steptoe announces the moment of successful embryo culture with that visually powerful double look—up to the moon and stars, down (and in) to the blastocysts just beginning their "journey towards life." I suggest that the link made by Sir Ian Lloyd between new scientific technologies for looking and the new reproductive technologies embodied in embryo research is anything but accidental. Rather, it reflects a deep connection in our cultural unconscious between the act of looking (whose object is, unconsciously, a woman's body) and the act of impregnating (which takes the same woman's body as its object). Such a connection not only predates, but actually may have shaped, the development of reproductive technology.

How does an understanding of the historical linkage of RT/VT illuminate the issue with which I began: the disturbing disjunction between mother and fetus, which many contemporary feminist theorists attribute to the effects of new postmodern visualization technologies? First, it reveals the modern sources of this disjunction: the RT/VT link is *modern* as well as postmodern in its origins and meanings. A similar RT/VT connection appears in the writings of Aldous Huxley in the 1920s through the 1940s: it was Huxley's work that inspired first Rock and Hertig, and then Edwards and Steptoe, in their accomplishment of in vitro fertilization.[75] And the 1930s films of Julian Huxley reflect—though with reversed emphasis—the connection between reproduction (bird or human) and another visualization technology, cinematography. If we return to the elements of cinematography as discussed by the Huxleys' close friend, Robert Nichols, we can locate in the modern moment an explanation for mother-fetus disjunction.

Nichols's second principle of the cinema is the satisfaction of desire through the mechanics of the camera; his third principle is what I have called the camera lie: "The peculiarity and beauty of the camera is that it can lie, and keep on lying." These two principles—the satisfaction of desire and the camera lie—are the essential link between visualization technologies and reproductive technologies. They constitute the fundamental essence not just of the modern cinema, but of the early modern microscope and the postmodern computerized Hubble telescope (named after Aldous Huxley's close friend) invoked by Ian Lloyd in the embryo research bill debate. In particular, that lie characterizes the endoscope and the laparoscope, central to the development of IVF, by Patrick Steptoe's own account. As Catherine Vasseleu has observed, these instruments not only represent, but they also distort, in the act of satisfying desire:

> In the endoscope's action of screening an image, can be read a desire
> for purchase reminiscent of the camera obscura, whose images not
> only entertained but put things in perspective. . . . Scientific images
> are not merely a form of photo-graphy; they participate in vita-
> graphy. . . . For a woman patient in an IVF clinic, the ovum being visu-
> alized on the screen, and collected with the assistance of laparoscopy,
> has all the reality of a metonymic assimilation. It is becoming part of
> the text which is her body.[76]

The desire that these instruments both serve and (re)enforce is, I would argue, precisely to see (and identify with) the fetus *separate from the body of the gestating woman.*

From Aldous Huxley's vision of the spermatozoa floating in an interstellar uterine space, no maternal body to be found, to his image of the babies in bottles that (as he told Anita Loos) most powerfully express the "triumph of science over nature," the literary articulation of this desire parallels its scientific articulation in the germinal and scopophilic obsessions of Edwards and Steptoe.[77] Awareness of this link between RT and VT can help us to arrive at a more flexible feminist response to both the mother-fetus disjunction, and to the technologies that—since the early years of this century—have worked together to produce it in representation and (gradually) in reality. It can encourage us to reinscribe the link between fetus and mother, and to remember that the embryo is neither astronaut nor "froward Homunculus," but part of a complex mutual relationship, beginning at conception and culminating, if not ending, with birth.

five

From Guinea Pigs to Clone Mums

NAOMI MITCHISON'S PARABLES
OF FEMINIST SCIENCE

> Have I any reason to think I was a budding scientist? Not really. . . . If I was anything, it was a good observer, not only of guinea pigs and wild plants, but also of people, and especially of people in relation to myself.
> —NAOMI MITCHISON, 1988

> The language of biology participates in other kinds of languages and reproduces that cultural sedimentation in the objects it purports to discover and neutrally describe.
> —JUDY BUTLER, *Gender Trouble*

Who becomes a scientist? What do scientists do? And what relation exists or should exist between the subjects and objects of scientific knowledge? Although (or, as we shall see, because) she did not define herself as a scientist, Naomi Mitchison's fiction has made an acute, and powerful contribution to the feminist critique of science. As Sandra Harding has comprehensively mapped it, this field of feminist science criticism has taken on a variety of projects: (1) equity studies; (2) studies of scientific "uses and abuses" in the service of oppressive political programs; (3) challenges to the notion of "pure science" itself, including the supposed value neutrality of the problems science chooses to study, the design of research programs, and the interpretation of scientific data; (4) studies of the discursive construction of the scientific project itself (exploring "the hidden symbolic and structural agendas" of science read as a text); and finally (5) the attempt to generate alternative feminist epistemologies from which a new science could spring.[1] Naomi Mitchison made major contributions to two of these projects

168

through her fictional challenge to "pure science," with its commitment to notions of progress and objectivity, and her representation of alternative feminist practices and epistemologies that could ground a new science.

Mitchison addressed these concerns in what I call her parables of feminist science: fictions that articulate her challenge to the gendered boundaries of modern science, as well as to the construction of the sexed subjects and objects of scientific knowledge. I understand these stories of science, sex, and gender as parables, not simply allegories, because they teach a lesson that is, in the broadest sense, *moral*. That is, Mitchison's dramatic and fictional parables articulate a feminist moral analysis of how human beings either do relate, or ideally should relate, to each other and to other living species.[2]

Western science has customarily been used to naturalize inequalities of race, sex, culture, even species.[3] Since the eighteenth century, the scientific intervention in reproduction has furthered that process of naturalization, as reproduction has come to function as a preeminent site for the negotiation of boundaries between self and other.[4] Reproduction has a moral dimension, too; inevitably, it raises issues of the relationship between self and other, mind and body, the limits of such concepts as autonomy, responsibility, control.[5] In fictions written over the course of seventy years, Naomi Haldane Mitchison has explored these dimensions of reproduction, telling stories about reproductive interventions ranging from sex selection and eugenics, to interspecies reproduction, parthenogenesis, grafting, cloning, and genetic engineering.

A crucial aspect of the work of feminist critics of science has been the search for "alternative visions and practices present within our own Western traditions, even within the domain of modern science."[6] Mitchison's play, *Saunes Bairos: A Study in Recurrence* (1913), and her science fiction novels *Memoirs of a Spacewoman* (1962; pub. 1976) and *Solution Three* (1973; pub. 1975) address the scientific manipulation of reproduction, both to articulate a pointed critique of that use of science and to embody some new visions of the relation between science and society and some new models for scientific practice. In a sequence of diverse scientific parables, figuring high priests, echinoderms, a parthenogenetic child, an alien graft, surrogate mothers of cloned fetuses or "clone mums," and the genetic code contained in DNA, these works challenge our customary understanding of scientists, experimental subjects, and the meaning and value of scientific intervention. As a background for those revisionary parables of alternative science, let us begin with a parable about guinea pigs—a parable that is not fiction, but fact, figuring not science-as-it-might-be, but science as it has been.

GUINEA PIGS

In the early years of this century, a little girl and her older brother shared a passionate interest in guinea-pig breeding. In a corner of their Oxford garden, where friends might have expected tennis courts or croquet fields, they installed some three hundred guinea pigs behind wire fences. Together, they studied the animals' behaviors, oversaw the births of countless litters, and charted the guinea pigs' changing coat colors and patterns from generation to generation. The childhood hobby led, for the boy, to his first scientific paper, "The Comparative Morphology of the Germ Plasm," and from there to an adult career as a geneticist.[7] The little girl grew up to be a novelist, who in a memoir would look back on her childhood and wonder, "What was my love-hate relation with science?"[8]

This tale of guinea-pig breeding from the childhood of J.B.S. and Naomi Haldane is gleaned from Naomi's own memoirs, her brother's biography, and the writings of friends and family. I read it, polemically, as a parable of science, sex, and gender.[9] It dramatizes the social construction of science in action: the mapping practices by which the scientist and the object of scientific knowledge are differently constituted. Conventions of autobiographical and biographical narrative add another layer to the story, shaping how Ronald Clark, Jack Haldane's biographer, tells it, and how Naomi remembers it. Gendered as well as generic conventions construct the different roles of brother and sister in relation to their joint scientific project: we can see this if we look at a brief passage from Clark's biography of Haldane: "The animals had multiplied, naturally and quietly in guinea-pig fashion, and while Naomi had learned to distinguish their individual voices, her brother was studying them to confirm the laws of inheritance as laid down in Mendel's revolutionary work" (30). As Sandra Harding has observed, "'scientific' and 'masculine' are mutually reinforcing cultural constructs"; this passage articulates that culturally constructed, gendered nature of science.[10] Clark's formulation embodies and deploys the distinctions between natural and constructed, sex and gender, animal and human, individual and collective, that come into play whenever the boundaries of science are policed. As this aspect of what I am calling the guinea-pig parable dramatizes, these distinctions enclose Jack Haldane and close out his sister Naomi.

If we turn to Naomi's memoirs, we can see how the children's different responses to the guinea pigs reflect their different positionings, even at that early age, in relation to the scientific project. Of course their guinea-pig breeding was as much play as scientific inquiry, for they were both still children. But play itself, as D. W. Winnicott has taught us, helps to shape what we understand as reality, both of psyche and society, and in Naomi's memoirs we can see the children constructing reality differently through their different ways of playing.[11] In *All Change Here*, Naomi says little about Jack's work with the guinea pigs except to detail how he directed their experimental progress—"Jack suggested that we should do serious genetics with them."[12] In contrast, she recalls her own absorption with the guinea pigs in detail, ambivalently labeling it "semi-scientific":

> I had started by keeping a few [guinea pigs] and gradually began to study them in a semi-scientific sense, listening to, identifying and copying their various squeaks and chitters, and seeing their relationship with one another and the whole pattern of guinea pig likes and dislikes. I even started taking notes. I certainly anthropomorphised too much, but I don't suppose any one else has ever watched guinea pigs in this way. Scientifically they are an exploited race. (61)

Two different models for scientific practice help to illuminate the significance of the difference in Jack and Naomi's relations, as protoscientists, to their objects of study. Evelyn Fox Keller has delineated a shift in scientific practices, both historically produced and gendered, between observation or representation (characteristic of nineteenth-century naturalism) and an intervention "that promises effective mastery over the processes of making and remaking life" (characteristic of twentieth-century experimental biology).[13] That distinction seems to apply here. Naomi herself invokes it (in the epigraph for this chapter) when she labels herself less a "budding scientist" than a "good observer . . . not only of guinea pigs and wild plants, but also of people." Thus Naomi worked at "identifying and copying [the] various squeaks and chitters [of the guinea pigs] and seeing their relationship with one another," but it was "Jack [who] suggested that we should do serious genetics with them" (61). Naomi's reliance on observation resembles the method of nineteenth-century biology, which devoted itself to "captur[ing] the mysteries of nature by documentation and description, rather than by a priori explanation," but Jack anticipates the twentieth-century trend toward experimentation.[14] As we shall see, this distinction between intervention and observation reappears in Mitchison's reproductive fictions, where a series of alternative visions of scientific practice suggests that the methodological boundaries of science, like gender boundaries, are not invariant and natural but rather culturally constructed.

We can also read this guinea-pig parable as part of a broader narrative of child development, during which gender-role demarcations are mapped onto both professional and personal life. Carol Gilligan's work on women's moral development can illuminate the connections between this story of guinea-pig breeding and that more general developmental narrative, by giving us an additional way of understanding the brother's and sister's different relations to their pets and/or experimental subjects. Gilligan's distinction in social positionings between the more typically female attention to a network of caring and responsibility and the more typically male focus on a realm of abstract principles is intended to describe relations between human beings, but it is remarkably apt for the human/animal relations of the children's guinea-pig breeding. Naomi responds to the guinea pigs as a network of individuals with whom she is in relation, but Jack considers them in the aggregate, as a collectivity about which he hopes to be able to formulate abstract principles that will ultimately produce an advance in scientific knowledge. He sees them as the collective illustration of a biological process ("the laws of inheritance"), but she sees them as communicative beings. She teaches herself to distinguish between individual guinea pigs, tries to learn their language in order to communicate with them, and is willing to breach social and biological taboos in order to learn about them—observing their matings, and even tasting their milk.[15] In short, Jack expresses the young man's greater interest in abstract (in this case, scientific) principles, and Naomi's attention to the individual experiences, language, and likes/dislikes of the guinea pigs seems to spring from what can be described as a distinctively female moral concern, expressed most directly in her observation, "Scientifically they are an exploited race." The parallel to women is unstated, but powerfully implied.

Not only Naomi's moral development, but her physiological development too involved gender-role negotiations that are expressed in this parable of guinea-pig breeding. Clark describes the guinea pigs' reproductive lives as proceeding "naturally and quietly." Although such a description hardly seems apt for the controlled and manipulated breeding schedule required of Mendelian genetics experiments such as the ones in which Naomi and Jack were involved, the guinea pigs do still function as the natural ground for those scientific interventions. In contrast, this guinea-pig parable reveals how Naomi's maturational process, from "boy into girl" and girl into woman, as she puts it in *All Change Here*, was shaped and determined by a variety of spoken and unspoken cultural codes far more constraining for women than for men (9). Her brother's scientific engagement continued unchecked into his adulthood, but her impending womanhood began to conflict with her interest in science.

In her memoir, Mitchison questions the extent to which her girlhood limited her access to the rich intellectual world of the laboratory. Although she remembers always having "a healthy respect for scientific curiosity and work," she felt she lacked her brother's "early understanding of it." And from the vantage point of maturity, she wondered whether this was a temperamental difference between them, or "whether certain avenues of understanding were closed to me by what was considered suitable or unsuitable for a little girl. Not deliberately closed, I think, since both my parents believed in feminine emancipation, but—there is a difference between theory and practice."[16]

Her parents' ambivalent attitude toward science for women trapped her in a gendered double bind: "Both parents encouraged scientific curiosity," she recalled in "Small Talk . . . ," "though there was a rule that we did not talk about anything below the diaphragm at meals" (99). Mitchison's scientific involvement with the guinea pigs was differently inflected for her, because of her impending womanhood. Curiosity about those regions below the diaphragm meant something very different for girls than for boys, as one memory in particular demonstrates. There was, Mitchison remembers in *All Change Here*, "one terrible blameword: *suggestive*": "I remember watching the male guinea pigs agitating around the females with their very typical chittering (and these were important matings from the standpoint of experimental genetics) and quoting, 'In the Spring a young man's fancy Lightly turns to thoughts of love.' I was heavily jumped on for saying something suggestive and felt deeply ashamed for weeks" (46). Reproduction is the site here of a boundary dispute. Although reproductive and sexual behavior is crucial terrain for scientific investigation, it is marked off-limits for Naomi because her study of such a scientific question would threaten her gendered social position as a modest, chaste, and hence marriageable young woman. If parental disapproval taught her that a scientific interest in reproduction is not gender-appropriate behavior for girls, it is little wonder that Naomi came to identify not with scientists, but with their objects of study: the guinea pigs.

This identification with her guinea pigs is explicit—and explicitly subversive—in her memoir *As It Was*. She tells how she sampled the milk "from the teats of one of my loved ones" but she named "every guinea pig appropriately, many after the famous scientists of the day" (61–62). Transposing experimenter and experimental subject, Naomi's naming play turns science subversively inside out. But interspecies identification, like interspecies analogy, has its limits. Although, as Mitchison remembers wryly in *All Change Here*, "I understood [guinea pig] sex life, with its genuine likes and dislikes, so well that I thought there was nothing about sex I didn't know," she learned in her own marriage that there was a crucial difference between guinea pig and human sexuality (61–62). The former operates by instinct (albeit shaped here by the controlling interventions of the scientific breeder), but the latter is constructed by powerful cultural forces.

Raised in a society that proscribed sexual curiosity in women (and to some extent in young men as well), Naomi found when she married at the age of eighteen that she had to unlearn some powerfully negative sexual habits, and in effect teach herself to respond sexually.[17] Both she and her husband Richard Mitchison were virgins when they married, with "astonishingly little idea of what [being in love] meant in marital practice." Intercourse was initially unsatisfying: "The final act left me on edge and uncomfortable. Why was it so unlike Swinburne? Where were the raptures and roses? Was it going to be like this all my life?" she wondered.[18]

Not until after the birth of her first child did sexual experience improve for Naomi, and then not because of any "natural" increase in sexual skill accompanying motherhood, but because of the intervention of a prominent birth-control educator and authority among modern feminists—Marie Stopes. The publication of Stopes's landmark work, *Married Love*, with its specific information on sexual techniques, in particular the importance of foreplay and the nature of the female orgasm, set in motion Mitchison's process of sexual self-education that resulted in "a marked increase in happiness."[19] The notion of a pre-cultural, natural sexuality—legacy of her experience with the guinea pigs, when she watched the creation of litter after litter and assumed that "there was nothing about sex that I didn't know"—gave way under the powerful impact of Marie Stopes's feminist sexology, to a lifelong, sophisticated investigation into the cultural construction of sexuality and reproduction.

This guinea-pig parable illustrates the cultural construction of science in a number of ways: scientific investigation is gendered male, thus excluding women from scientific practices; animal reproductive practices are constructed as a natural ground for scientific interventions; human behavior is reconceptualized under the category not of nature but of culture, as "scientific experts" intervene to shape the contours of human sexual behavior. But it also provides glimpses of the ways that Mitchison would subvert these acts of scientific boundary-making: her transgressive naming of the guinea pigs after the great scientists whose experimental objects they were; her cross-species identification with the guinea pigs' sexual and maternal experiences; and most striking perhaps (because contradictory to that identification) her acknowledgment that there are epistemological limits to science, and her suggestion that the interpretive categories of science may themselves require revision.[20] Like this parable of fact, a series of fictional parables demonstrates that Mitchison finds alternative scientific possibilities in a familiar site of scientific intervention: the reproductive body, whether of a guinea pig, an extraterrestrial creature, or a woman.

HIGH PRIESTS

Enthusiasm for eugenics grew steadily in the early years of the twentieth century, following the rediscovery of Mendelian genetics. In 1909, the Eugenics Education Society was founded, by statistician and Victorian amateur scientist Francis Galton, close friend of Thomas Henry Huxley. Two years later, the Oxford University Union affirmed eugenic principles by nearly a two-to-one vote, and in Cambridge, Kevles recounts, Eugenics Society meetings were crowded with hundreds of students, dons, and townspeople (60). By the time young Naomi Haldane began writing her play about eugenics, *Saunes Bairos: A Study in Recurrence* (1913), Britain was in a eugenic fervor, fueled both by anxiety over perceived racial degeneration and by hopes of using prenatal culture to produce racial perfection. In 1913, the year the play had its premiere at the Oxford Preparatory School, "a London woman, pregnant and enterprisingly Lamarckian, betook herself to plays and concerts, conversed with H. G. Wells among other writers, and . . . gave birth to 'Eugenette Bolce,' who was widely hailed as England's first Eugenic baby."[21]

Although there is no record that Eugenette Bolce's mother-to-be traveled to Oxford to swell the audience when *Saunes Bairos* was premiered, Naomi Haldane's mother and father were there, along with a large section of the white, male British scientific and professional elite—Oxford dons, local professional men—and their families. Moreover, the cast was drawn from the children of the intelligentsia: its director was Lewis Gielgud, Naomi's lifelong close friend, and the players included J.B.S. Haldane, Aldous and Trevenen Huxley, Richard Mitchison, later to become Naomi's husband, and Naomi Haldane herself as the heroine, Carila. (Ill. 17.) Many of those family members and close friends would produce significant challenges to the eugenic project. In particular, her brother J.B.S. Haldane, having learned his Mendelian genetics by her side as a fellow guinea-pig breeder, would become one of the most vigorous critics of mainline eugenics, along with his close friend Julian Huxley. *Saunes Bairos* took the mark of those people and events around her, but it also reflected her own specific experiences as a young woman in early-twentieth-century Oxford.

Set in the Andes around the year 1200, and in present-day British Guiana, *Saunes Bairos* challenges the eugenicists' cherished notion of scientifically produced reproductive progress, dramatizing as an object lesson for contemporary society the downfall of a country led by High Priests who have made genetic science into a religion. The rulers of the Andean country of Saunes Bairos have applied their understanding of the principles of genetic control, learned through their traditional practice of guinea-pig breeding, to the creation of a eugenic program of human reproductive control aimed at producing the perfect race. Transcending its costume drama format, the play registers a stinging critique of eugenics from the perspective of women and third-world peoples. *Saunes Bairos* bravely calls into question the foundational beliefs of its scientifically informed, elite audience: Louisa Trotter Haldane's imperialism and the scientific practices of reproductive control intimately linked to Naomi's family and friends.[22] But more than its willingness to take on issues of racism, colonialism, and scientific control of sexuality and reproduction, *Saunes Bairos* is notable for the perspective from which those issues are raised: one more sympathetic to women and people of color than to the white, Western scientists who made up such a large part of its audience. In its deliberate adoption of a marginal perspective in order to

17 *Saunes Bairos:* The Cast
Insets:—H. K. Ward, Miss S. Egerton, W.G.K. Boswell.
1st Row:—P. Leigh-Smith, G.L.M. Clauson, A. R. Herron, R. Hartley,
G. C. Vassall, A. Huxley, F.A.H. Pitman, D. Mackinnon, D. N. Hossie.
2nd Row:—H.E.E. Williams, Miss E. Blockey, F. H. Brabant, H. A.
Robertson, S.S.G. Leeson, A. Egerton, G. R. Mitchinson, P. Hartill, N. T.
Huxley, Miss L. Petersen, W. E. Hayes, D. Fraser, D. Hardman.
3rd Row.—Miss E. Hayes, Miss B. Gray, Mrs. Haldane, J. D. Denniston,
Miss M. Fraser-Tytler, Miss N. Haldane, L. Gielgud, J.B.S. Haldane,
Miss K. Richards, N. P. Birley, Miss E. Collier, Miss F. Petersen.
4th Row:—Miss H. Cooke, A. Barker, Miss M. Cooke, R. Barton,
Hon. C. C. Farrer, J. D. Walker, T. Greenidge, Miss G. Petersen,
Miss G. Berkeley, K. Jefferson, U. Pentreath.

critique mainstream scientific policies, this play established the critical
voice and position that would characterize Mitchison throughout her
writing life, enabling her to find a fresh perspective on scientific values
and practices.

The monitory thesis of *Saunes Bairos: A Study in Recurrence*, is tele-
graphed in its subtitle: the play recounts a scheme for perfect genetic
control gone awry, a scheme that could recur. Sir Ian Lloyd, speaking
in the Human Fertilisation and Embryology Bill debate in 1990, observed: "It was once said of Spinoza that he was drunk with God.

There is, I accept, an equal danger that modern man can become drunk with science. If I am compelled to choose, as we are by this measure [the Human Fertilisation and Embryology Bill] tonight, I have no doubt which form of inebriation I prefer—even if, as I suspect—they are not mutually exclusive."[23] In *Saunes Bairos*, Naomi Haldane anticipated Lloyd's point by nearly eighty years. She warns of a new breed of High Priests subject to a new form of intoxication: not God but the temptation to play God, as the revelations of genetic science lead to a fantasy of perfect eugenic control.

The horizon of Mitchison's play is just such an attempt at genetic control, waged by the High Priests of Saunes Bairos. We learn in the prologue that they caused the downfall of their small Andean nation by creating a law forbidding the citizenry to have more than two children, one boy and one girl: "First the Law was above all, and by the mouth of the High Priest it elected how many children any man might have, and for what qualities."[24] This Law, laid down ostensibly to limit the population of this landlocked, isolated South American country to a sustainable number, really served the Priests' dreams of world conquest. They planned that "a perfect race should be evolved—perfect in mind and body—so that by and by the Priests should lead them out of Saunes Bairos to conquer the world" (2). Naomi's protagonists, the nonidentical twins, Carila and Coraxi, determine to test the sacred Law that regulates and limits birth and forbids citizens from crossing the snow line to leave their native valley.

The play's focus on a brother-sister pair highlights gender-based differences in ways that recall Naomi's own childhood with her brother Jack. But here, the difference lies not in their different access to science, but in their different relation to religion, specifically in their different levels of resistance to Priestly patriarchal law. As in the contrast between Naomi's and Jack's responses to their guinea pigs, here too we have an opposition between woman's commitment to a network of caring and responsibility, and man's allegiance to abstract principles. Carila's resistance is based on the subversive reframing of the Law taught her by her old nurse, who in turn learned it from her mother: "The Law was made for the people, not the people for the Law" (3).[25] Crossing the mountains to escape the Law, Carila hopes to find a better world for her people. Yet her brother Coraxi puts the Law before the people, abstract reasoning before relationships. Motivated by masculine rivalry, he means to protect Carila from "this crew of men that she's picked up" and he clings to the confused conviction that "one can cross the mountains in a reverent spirit towards them and the greater principles of the Law which they suggest" (15). The failure of Coraxi's plan dramatizes the fact that freedom from tyranny of the fathers cannot be obtained through resistance in the name of fraternity. Both modern fraternal patriarchy and classical priestly patriarchy preserve intact the sexual contract that relegates women to the private sphere and constructs reproduction as something to be regulated by society.[26]

Brother and sister respond differently to the challenge they must meet if they want to escape Saunes Bairos: to be permitted beyond the snow line, they must learn the Priests' secret password. Trusting in her own beliefs, questioning the authority of the priesthood, and committed to her comrades, including the blind and the lame, Carila overhears the password and uses it successfully to escape Saunes Bairos. But deferential to his priestly *and* biological fathers, scornful of those physically different from himself (the blind, the lame, and women), defensively insistent on his own masculine superiority, Coraxi is unable to trust the phrase his sister has overheard. While she and her band of followers escape, he is trapped on the wrong side of the snow-line. The best of his followers are killed; the worst remain in Saunes Bairos, condemned to the reign of a new, but equally oppressive, group of High Priests.

The contemporary implications of the ancient story of Saunes Bairos are articulated in the play's framing scenes, set in present-day British Guiana. In a market-day encounter over guinea pigs, Naomi presses home the moral: Western scientists risk producing another Saunes Bairos if they do not rethink their commitment to eugenics. The Englishmen boast of their plans to apply eugenic principles on a large scale, and the Guianian guinea-pig sellers are horror struck at the prospect of "[doing] again all that was done in Saunes Bairos, the country which was governed by the Law, and because of the Law, fell" (2). As the Guianian guinea-pig seller explains to the English gentlemen who are their interlocutors: " . . . because the Law did not grow with the minds of the people, things came to pass . . . and behold now the Law was dead, and it is being born again in other lands than its own" (3).

Saunes Bairos demonstrates scientific sophistication impressive in a fifteen-year-old—in its theme of eugenics, and its allusions to the physical anthropology of Franz Boas, the biometrics of Karl Pearson, and to the Mental Deficiency Bill passed by the government in 1913.[27] Yet its most remarkable contribution lies not in its scientificity, but in its use of the parable of the High Priests to articulate a critique of the elitist and hierarchical model of science. Asserting that there are—and that there should be—limits to science just as there should be limits to religion, Naomi Haldane's play cautions us that science, like other Laws, must "grow with the minds of the people" rather than being held in secrecy by an elite (3). *Saunes Bairos* challenges the notion of a scientific priesthood, affirming instead a model of scientific practice that would be picked up by feminists from Vera Brittain to Donna Haraway. Presenting an alternative vision of Western science from the perspective of women and South American peasants, *Saunes Bairos* suggests that unless science is responsive to the needs of its community, subject to local conditions, and the product of situated knowledges, its trajectory will be not emancipatory and progressive, but oppressive and regressive.[28]

ECHINODERMS

"One cannot imagine oneself a starfish or a sea-urchin however hard one exerts one's mind," wrote Julian Huxley in his coauthored biology textbook, *The Science of Life*.[29] In *Memoirs of a Spacewoman* (1976) Mitchison challenges Huxley's dogmatic assertion that there are distinct limits to the empathetic and epistemological reach of the human species.[30] Beginning with precisely the act Huxley says is impossible— imagining oneself a starfish or a sea urchin—Mitchison shows how that imaginative act radically transforms the notion of scientific objectivity. From a refusal to allow individual feelings for or about the experimental subject to color the interpretation of the data, Mitchison's model of science moves to a collective acknowledgment that such individual feelings exist and must be addressed by the critical scientific community.[31] Another way to describe this shift might be from the notion of a science purged of all subjective feelings, to a science that acknowledges, builds in, and corrects for, the existence of subjective feelings. Read as a parable, Mitchison's starfish narrative dramatizes how a vision of science as social knowledge can shape the new scientific culture invoked by Andrew Ross in *Strange Weather*: the kind of culture that "can learn from differences," rather than recoiling from them or suppressing them in fear (12).

The premise for *Memoirs of a Spacewoman* is the notion that biology has undergone a methodological revolution: its goal now is not to achieve control over new territory, but to establish communication with new species. The novel's protagonist is a scientist, Mary, who as a "communications expert" on an interplanetary zoological expedition, must interact—without interfering—with the inhabitants of Lambda 771. They are beings whose "evolutionary descent had been from a radial form, something like a five-armed starfish" (20). Mary's scientific task reflects a different construction of the aim of biology: not to achieve knowledge of (and consequently power over) species understood as ineluctably alien, but rather to attain communication with alien species without trying to turn them into kin.[32] The biological method is no longer scientific objectivity, but what one might call scientific empathy: the strategic assumption of the position of the other, as a first step in establishing communication. In order to make contact with the radiates, Mary must participate in their phenomenological world—must imagine herself one of them. To her surprise, she is so successful at the task that Huxley had declared impossible—imagining oneself an echinoderm—that she barely makes it back into human consciousness. *Memoirs of a Spacewoman* is a sustained meditation on the necessary risks of such a reformed scientific method.

Mitchison's novel is grounded on a painful fact: communication without interference is an impossibility in our post-Heisenbergian world. Interference is inevitable, because any communication inevitably changes both (or all) parties in the interaction.[33] But Mitchison digs deeper even than the interference that comes with communication, examining the impact of the imaginative act of empathy that, in *Spacewoman*, is the prior condition for communication. In her starfish parable, Mitchison suggests that even to identify with the object of scientific knowledge can remap the boundaries of scientific investigation, reshaping the kinds of questions a scientist asks, the relation between the subjects and objects of scientific knowledge, even scientific practices themselves.

What would it mean to imagine oneself a starfish or a sea urchin? When the episode begins, Mary's musings locate her firmly within the binary discourse of modern science. She relies on a favorite example of science writers: the echinoderm. "It is only in circumstances like this that we realize how much we ourselves are constructed bi-laterally on either-or principles. Fish rather than echinoderms" (20). There is an intertextual history to such references, as the epigraph to this section suggests. Thirty years before Mitchison wrote *Memoirs of a Spacewoman*, her good friend Julian Huxley used an echinoderm in the textbook he coauthored with H. G. Wells and G. P. Wells, *The Science of Life*, to illustrate an otherness so profound that the human imagination cannot compass it:

> Most of the animals that we have considered so far bear a certain very
> crude resemblance to ourselves in the way their bodies are laid
> out . . . for the most part, by stretching one's imagination just about as
> far as it will go, one can get a crude idea of what the animals feel
> like. . . . But with the phylum *Echinodermata* things are different; with
> a few exceptions they do not know front and back, right and left. They
> are not bilaterally symmetrical, as we are. They do not "look before
> and after." One cannot imagine oneself a starfish or a sea-urchin how-
> ever hard one exerts one's mind. (222)

Naomi Mitchison was familiar with Huxley's textbooks, even crediting
one of them in *All Change Here* for helping her get through science pre-
lims in wartime Oxford (116). Even if she had not read *The Science of
Life* for years when she began to write *Memoirs of a Spacewoman*, still she
would have been familiar with the orthodox scientific position that the
textbook served to promulgate. Thomas Kuhn has argued that scientific textbooks function principally to consolidate, and normalize, the
gains of revolutionary science.[34] As Kuhn's argument makes clear, the
very fact that echinoderms functioned as an illustration in the textbook signals the nonrevolutionary nature of the intervention. Then
why should echinoderms be so prominent in a novel challenging the
construction of modern life science, or "Terran" science as Donna Haraway has phrased it? What previous site of controversy or contestation was normalized by the textbook reference to the echinoderm?

Like the guinea pig, the echinoderm was a prominent object of scientific study in the late nineteenth and early twentieth centuries. A life
form that has played a crucial role in the modern conceptualization of
reproductive mechanisms, echinoderms were used to produce knowledge about reproduction that has then been applied to human beings.
As studied by Hans Driesch and Jacques Loeb, the echinoderm played
a central part in two major findings in reproductive biology, both of
which Mitchison dramatizes in her fictions: the decisive solution to the
debate over preformation or epigenesis (which returns as a central
theme in *Solution Three*) and the first instance of laboratory-induced
parthenogenesis—the development of an embryo from an unfertilized egg—which gives rise to an episode in *Memoirs of a Spacewoman*.

The question of preformation or epigenesis was one of the basic
riddles of embryology: whether the individual was already preformed
in one of the gametes, or developed from lesser to greater organization in the course of gestation. As Gordon Taylor summarizes this controversy in the textbook he titled, in clear homage to Huxley et al., *The
Science of Life*: "Is the adult contained with all its parts pre-formed in

the egg, or is it put together from raw materials?"[35] Hans Driesch's experiment on sea-urchin eggs in 1891 clarified the developmental relation of heredity and environment and thus provided the decisive solution to this age-old riddle. Mechanically separating two-celled sea-urchin embryos, he allowed each half-embryo to develop for a day, and found "that each had formed a perfect sea-urchin larva, somewhat smaller than usual."[36] That separated cells produced not a partial embryo, but one that was complete and entire, if somewhat smaller than usual, suggested to Driesch that cells had the capacity to respond to their environment, redirecting growth as the environmental context required. The echinoderm experiment revealed to Driesch that "differentiation might be the result of cellular responses to both internal and external conditions."[37] Driesch's experiment thus advanced the scientific understanding of embryonic differentiation, the same principle that had fascinated Julian Huxley from his childhood reading of *The Water-Babies* to his adult work with the axolotl.

The scientific understanding of embryonic development was given an even greater jolt in 1901, when Jacques Loeb developed a technique for parthenogenesis through the mechanical stimulation of sea-urchin eggs. Loeb demonstrated that "he could cause an unfertilized sea urchin egg to undergo development simply by pricking it with a needle or changing the salt concentration of the seawater in which it was being cultured."[38] This experiment caused great excitement in both the scientific and the popular press: "To laymen, the successful production of parthenogenetic organisms was both exciting and terrifying, presaging the era in which life would be created in a test tube."[39] The echinoderm's crucial role in modern embryology (in particular in the work of Loeb and Driesch) suggests why it is so central to Mitchison's *Memoirs of a Spacewoman*: it serves as a powerful image for the biological and social implications of scientific intervention in reproduction, on a genetic and/or an environmental level.

The echinoderm embodies unbridgeable difference—both of body and of psyche—as is suggested by the rather fanciful description of it in *The Science of Life*. Huxley returns to his touchstone text, Kingsley's *Water-Babies*, to describe the different epistemology of the echinoderm. Starfish and sea urchins not only have radial rather than bilateral bodily structures, but they don't discriminate between right and left, front and back; they don't "look before and after," as Huxley puts it. The phrasing recalls the two brothers of Mother Carey's story. "One was called Prometheus, because he always looked before him, and boasted that he was wise beforehand. The other was called

Epimetheus, because he always looked behind him, and he did not
boast at all, but said humbly . . . that he had sooner prophecy after the
event."[40] Neither Promethean nor Epimethean, the echinoderm has a
wholly different structure of consciousness.[41] The textbook distinction between echinoderm and human being normalizes the structural
principles of human consciousness or human epistemology, contrasting them to the epistemological world of the wholly alien, imaginatively ungraspable echinoderm.

To return to Mary's experience in *Memoirs of a Spacewoman* in which
she finds herself thinking like a radiate: because she has been nudged
by her imaginative identification with the radiates into a proto-
deconstructive questioning of the binary foundation of Western
thought, Mary finds herself virtually incapable of the binary thinking
fundamental to scientific investigation. An empathetic response (and
on the part of a scientist, at that) destabilizes for her not only the categories by which she understands herself and others, but also the even
more fundamental categories that she deploys to produce knowledge
about living beings and about our surroundings: geographical categories like up and down, front and back; temporal categories like past
and future; classificatory categories like good and evil, black and
white.

Scientific practices are built on the act of classification and differentiation. To imagine oneself a starfish or sea urchin is to transgress
those classifications, to cease to be objective, both about oneself *and*
about the echinoderms. As Mary experiences the empathy that is the
necessary first step in establishing communication with this alien life-
form, she begins to imagine life from their perspective, with dramatic
results:

> One is so used to a two-sided brain, two eyes, two ears, and so on that
> one takes the whole thing and all that stems from it for granted. Incor-
> rectly, but inevitably. My radiates had an entirely different outlook. As
> I got to know them better, I realised that in many ways they were
> highly—in our phrase—"civilised." But they never thought in terms of
> either-or. It began to seem to me very peculiar that I should do so my-
> self, and that so many of my judgements were paired; good and evil,
> black or white, to be or not to be. Even while one admitted that moral
> and intellectual judgements were shifting and temporary, they had still
> seemed to exist. Above all, judgements of scientific precision. But after
> a certain amount of communication with the radiates all this smudged
> out. (Mitchison, *Memoirs of a Spacewoman*, 26–27)

Released from binary physiology and its psychic extension, either/or thinking, Mary enters into a wholly new experience of her body and, by extension, a new philosophy. "I was beginning to think of general philosophical problems in a way that seemed new and full of possibilities" (28).

Between the assertion that "one cannot imagine oneself a starfish or a sea-urchin however hard one exerts one's mind," and Mary's experience of imaginative union with her radiate objects of study, lies a shift in the structure of scientific knowledge: from distanced objectivity to situated empathy, as well as a shift in the position of the subject of science, from center to margin.[42] That shift is a profoundly risky one, as Mary's experience goes on to reveal.

The conclusion to Mary's encounter with the radiates extends the implications of this parable to the realm of everyday life, suggesting that although polarized, so-called objective thinking is inadequate, both in life and in scientific practice, there are also risks associated with the multiple positionings born from empathy. Mary is attacked by a "jag," a large mosquito-like predator, and she freezes, mentally paralyzed by her empathy for the radiates. The episode suggests that a process of communal assessment and correction, or what Helen Longino has called "transformative interrogation," is essential to the new science relying on this method of situated empathy. Mitchison shows this in process when Mary is unable to snap back to human consciousness, and her fellow scientist and friend attempts to help her reorient herself. "Make a choice, Mary. A quick two-way choice," he urges her. "Shall we have a baby?" (30). Although Mary had already considered having a baby with her colleague, she now finds herself unable to answer: "I remember that I didn't answer. Couldn't. 'Baby or not baby?' he said. 'Two-way choice, Mary. Quick!' . . . And quick, I said to myself, quick, and he under his breath was saying the same. But I couldn't get back to myself. I couldn't speak. I couldn't say yes" (30).[43]

The impossible two-way choice, "Baby or not baby?" recalls one of Mitchison's own nonfiction interventions into reproductive practice: the 1930 pamphlet *Comments on Birth Control*.[44] Taking issue with the notion that scientific advances in contraception have been unproblematically emancipatory for women, Mitchison instead explores the complicated and ambivalent feelings that cluster around the issue of reproduction. "Apparently, all the feminist battles are gained, or almost all. Actually, nothing is settled, and the question of baby or not baby is at the bottom of almost everything" (32). Arguing that "intelligent and truly feminist women" want to "live as women" (that is, to satisfy the maternal drive by having many children and to attain sexual satisfaction) and to work, Mitchison holds that "adequate contraceptive methods are an essential part of this compromise" (25).[45] And yet, her point is, two-way choices are inadequate responses to the complexities of a woman's reproductive and sexual life. Such contraceptive methods do not completely solve the conflicts caused by the complex, multiple and contradictory desires that actual human beings actually have.

A willingness to acknowledge the limits of science and medicine links Mitchison's 1930 pamphlet to her 1976 novel. Her assessment of the inadequacies of birth control technology in *Comments on Birth Control* resembles her portrait, later in *Memoirs of a Spacewoman*, of Mary's experience with radial thinking, which teaches her the limits of the binary construction of experience basic to scientific understanding. In each case, it is the profound complexity of the seemingly simple reproductive choice—"Baby or not baby?"—-that grounds her critique of the dichotomous structure of scientific/technical thinking.

The attempt by Mary's fellow scientist to reorient her in the human bipolar epistemology embodies the kind of collaborative critique that is essential to the new science Mitchison embodies. But equally essential —and repeatedly asserted—is the value of the practice of identification that the colleague disrupts in the name of ongoing science. In the tension between the two positions lies the new, redefined, situated scientific objectivity that *Memoirs of a Spacewoman* dramatizes.

A PARTHENOGENETIC CHILD AND AN ALIEN GRAFT

Two other episodes in *Memoirs of a Spacewoman* can be read as parables that extend the theme of reproduction from the echinoderm to the human, offering an alternative perspective on the scientist as woman. In the first, Mary conceives and delivers a parthenogenetic child; in the second, she engages in a voluntary act of auto-experimentation, in which she permits her body to be used as a host for an alien graft. Although each episode would reward more extensive analysis, we need only sketch their outlines in order to examine how each functions as a parable or moral lesson. They situate a revisionary critique of science within a challenge to our conventional construction of the reproductive relationship between mother and fetus/infant, and prompt us to rethink our definitions of self and other, same and different.

In the first episode, Mary is stimulated to parthenogenetic conception by a Martian colleague. Realizing that the fetus she carries will be "haploid"—possessing only half the normal number of chromosomes—she considers aborting the fetus. It is certain to be sterile, and is likely also to be smaller than normal. But she decides to go ahead with the pregnancy for reasons that are, at the time, primarily scientific: "No doubt the thing could be stopped. . . . Yet would not this be interrupting an interesting and perhaps valuable experiment?" (64). Although her investment in the experience had been strictly scientific, once the child is born, the experience of mothering her small daughter moves her from detachment to engagement. Mary names her Viola, because the name brings to mind "not only a lovely, delicate and many-coloured flower and a musical instrument, but also as near a two-sexed person as we get on earth" (66).[46] Her love for Viola broadens her capacity to appreciate a range of differences (of gender, of size, of genetic makeup), while ironically limiting her sense of the determining power of such differences. This experience of mothering her haploid child convinces her that there are limits to her influence, as a mother, over the child that is born of her body.

In contrast, Mary's experience of voluntarily acting as host for the alien graft teaches her just the opposite: that she has underestimated the extent to which a creature seemingly separate from her can influence or shape her own experience—both mentally and physically. Although she begins the graft experiment with only the most detached, objective scientific interest, in the course of the project she becomes both psychologically and physically bonded to her graft. "As the graft grew, I began to have feelings of malaise, of the kind which one understands used to be common during pregnancy, though they are so no longer. . . . I began to be possessive about my graft; I could not think about it coldly" (53). In short, she is "becoming antiscientific" (148). The graft experiment verges on disaster, for Mary and all of the other hosts. It induces sweeping personality changes that threatened not just their status as scientific investigators, but as individuals, as Mary realizes when she considers the experiment in retrospect: "I had a good look at the very curious photographs which Pete had taken of me with my graft; it gave me peculiar residual feelings even now. Yes, I had been somebody else. Somebody, from a scientific point of view, delinquent" (159).

Part of a broader redefinition of the scientific project, the episodes of the haploid child and the alien graft can be read as parables or moral lessons. They offer new ways of thinking about pregnancy: as

an act of emotional rather than genetic connection in the case of Viola
the haploid child, and as a disturbingly invasive, identity-altering sym-
biosis, in the case of the alien graft. Expanding on the theme of multiple rather than binary approaches to knowledge introduced in the
echinoderm episode, these episodes assert the dangerous complexity
of communication, problematizing authoritative sites of scientific
truth. In the case of Viola, that authoritative site is genetics; we learn
that a genetic analysis does not comprehend all of her potential, for
she accomplishes an act of communication of which neither her
mother nor her Martian pseudo-father thought she was genetically capable. In the case of the graft, we see the limits of immunology, for the
graft's powerful effect on Mary's psyche demonstrates that an immu-
nological analysis of the interaction between host and graft cannot anticipate the range of psychological and environmental interactions
that the host will experience.

I began by arguing that Mitchison is best understood as a feminist
critic of science, whose fictions offer the vision of a new scientific culture, one which can tolerate and even learn from difference, one
which no longer accepts as adequate notions of progress and objectivity.[47] Reproductive variations function as points of entry to such alternative possibilities, in Mitchison's works. In *Saunes Bairos* she mines
the connections between contemporary eugenics and techniques of
population control practiced by a fictional Andean civilization to
trouble the notion of scientific progress. In *Memoirs of a Spacewoman*,
the parables of the echinoderms, the haploid child, and the alien graft
call into question the possibility of objectivity and the classic distinction between the subject and object of scientific knowledge. In *Solution
Three* (1975), she departs from the critique of masculinist Western science, to question instead the value of scientific intervention in the
name of an emancipatory agenda.

CLONE MUMS AND THE MASTER MOLECULE

What dictates the development of a human being: genetic makeup or
environmental influences? This nature/nurture controversy is a central theme of Naomi Mitchison's *Solution Three* or, as it was titled in
draft, "The Clone Mums."[48] Portraying a future society in which reproduction is carried out by cloning and surrogacy, heterosexuality is
stigmatized and homosexuality the norm, the novel raises a series of
questions crucial to the feminist critique of science: Is it acceptable to
exercise reproductive and sexual control in the name of a nonracist,

nonsexist, nonviolent agenda? What does it mean to privilege heredity over environment, genes over gestation? Can a society dedicated to scientific control reintroduce the possibility of individual and social choice?

That Mitchison's novel is dedicated "To Jim Watson, who first suggested this horrid idea," suggests that an allegiance to genetics and heredity is the novel's starting point.[49] Codiscoverer of the double helical structure of DNA, the molecule that transmits the genetic code governing human development and growth, James Watson with Francis Crick formulated the "central dogma" of DNA's role as "the master molecule."[50] Watson prefaced his autobiographical narrative of that scientific accomplishment, *The Double Helix* (written in 1967, three years before "The Clone Mums") with the notation, "For Naomi Mitchison."[51]

As these mutual dedications attest, a warm personal relationship existed between Mitchison and Watson, who was a school friend of her sons Avrion and Murdoch. Watson spent the Christmas of 1951 with the Mitchison family at Carradale, their country home in Scotland, soaking up the scientific conversation and (by his account) shivering in the unheated mansion. In *The Double Helix*, he recalls being cornered by Murdoch "to talk about how cells divide," and taking a long walk "with Av talking about his thesis experiments on the transplantation of immunity" (63, 66). Watson's affection for Naomi Mitchison was obviously colored by her sex and his sexism, for *The Double Helix* also mentions his memory of playing "intellectual games" under the "condescending stares of the Mitchison women" (64). It was not only in his social manner that Watson exhibited a nearly legendary misogyny, however; his scientific commitment to a notion of "the master molecule" expressed the gender bias in biology extending all the way back to Aristotle's theories of the passive egg as environment for the active and form-producing male sperm.[52]

If Mitchison's friendship with Watson in the 1950s, 1960s, and 1970s exposed her to sexist biological positions, another friend, the socialist embryologist C. H. Waddington, was instrumental in promulgating an opposed position within biology. As far back as the 1930s and 1940s, he located embryological determinism not in the cell nucleus (and its genetic material) alone, but in the interaction between the nucleus and the cytoplasm. Waddington, whose model for cellular interactions reflected the experiment in egalitarian living that he shared with his architect wife, "viewed the marriage of nucleus and cytoplasm as a partnership."[53] In his *Organisers and Genes* (1940), he attempted to demonstrate a balance of influence between components of the cell, "neither dominating the other. His cell, like his notion of marriage, was a partnership between equals."[54] Friendship aside, Waddington's scientific views would still have been well known to Naomi Mitchison, who in the jacket blurb to *Solution Three* describes herself as coming from "a family of scientists." Waddington was part of the cohort of socialist scientists that included her brother J.B.S. Haldane, crystallographer J. D. Bernal, and embryologist Joseph Needham.[55]

This early community of politically engaged scientific thinkers seems to have given Mitchison the resources to respond to Watson with more than simple approbation. As its plot unfolds, *Solution Three* reflects that more nuanced and complex view in several ways. In the foreword, Mitchison signals that she will be invoking the debate between these two different views of embryological determinism, when she takes the unusual step of providing her readers with definitions of two biological terms: *clone* and *meiosis*.

1. A clone consists of the descendants of an individual produced through a-sexual reproduction, having identical genetic constitution.
2. Meiosis is a cell division in the germ cell line during the formation of eggs and sperm, which results in the chromosome number being halved and the genes re-assorted. Note that, when egg and sperm cells re-unite in the process of fertilization, one chromosome in a pair comes from each parent, so that the original number of chromosomes comes back, but there may be crossing over of chromosome material. (6)

If we understand cloning as a reproductive technique producing control and sameness and meiosis as a reproductive event that functions randomly to produce variation and difference, we can see that these two biological terms can also serve as metaphors for social practices. The movement from enforcing cloning to permitting reproduction that has included meiosis (so-called natural reproduction) becomes, on a social level, a movement from control to chance, from the enforcement of sameness or identity to a toleration of difference. Although the narrative of *Solution Three* is biologically grounded, it moves from the exploration of different models for cellular structure and embryological development to a consideration of their implications for the social sphere. Like Charlotte Haldane's *Man's World*, Mitchison's *Solution Three* draws parallels between human biology and human society based on the understanding that neither realm is free from the imprint of ideology. We have seen that Haldane's future state is modeled on the cell; Mitchison's future state is governed by "the code," a term deliberately evocative of the chromosomal "code of codes" contained in the double helix of DNA.[56]

Exploring how gender relations are played out both in our scientific findings and their social application, Mitchison's novel works simultaneously as a direct intervention into the scientific understanding of reproduction and as a critique of the social and political uses of science. We must read this novel on two levels, then: the cellular and the social. We can map out this double-layered investigation as a parable (on both cellular and social levels) of the relation between the clone mums (for whom the novel was first titled) and the "master molecule" DNA, to whose discoverer the novel was first dedicated.

The plot of *Solution Three* can be summarized fairly quickly. The Council governing Mitchison's society of the future has responded to a disastrous period of war, overpopulation, food shortage, famine, and violence by adopting a policy known as the Code (13). This term, whose juridical/biological ambiguity Mitchison exploits, refers to a program of mandatory scientific and social controls on human sexuality and reproductive behavior.[57] Like Julian Huxley's "Tissue-Culture King," *Solution Three* posits the perfection of the technique of tissue culture, which makes it possible to preserve and extend the power of political leaders in a literal, biological sense. But unlike Dr. Hascombe's self-aggrandizing use of power in his African pharmacracy, the goal of the tissue-culture operation in Mitchison's novel is social emancipation. The black man from the United States and the white woman from the Shetland Islands whose political activism on behalf of disenfranchised peoples makes them the prototypes of the perfect human being were cloned, after their deaths, by the enlightened framers of this new world order. Their clone children are born to surrogate mothers selected by the state (called "clone mums") and deliberately conditioned after birth in a process ("the strengthening") intended to replicate the environmental forces that shaped the original He and She (59). These clone children are intended to form the population base of a world where sexism, racism, and violence are no more. The gene pool of this world is so drastically restricted (through cloning and the creation of a social stigma on so-called natural reproduction) that it has become nearly identical from generation to generation: social stability has been wrested from a period of turbulent, aggression-swept instability.[58]

Of course, the very existence of a social norm requires a corresponding realm of deviance, in Mitchison's society as in our own. The novel acutely dramatizes how the margin is both produced by, and produces, the center: "In fact the Cloning was due to Professorials. It could never have been done without the work of the biologists Quereshi and the great Sen and, earlier, Watson and Mitchison. But the Professorials had been unwilling to accept the inevitable new morality of Solution Three, essential to human survival, and, according to the code, force was not to be used on them. Only understanding" (16). "Understanding" is an insidious term for suasion by various social technologies. Rather than using force to gain compliance with the code, the governing council of this new society has relied instead on conditioning, social pressure, incentives for homosexuals and clones (such as preferential treatment in housing), and disincentives for heterosexuals and biological parents (such as social stigmatization and inadequate housing). Although they still engage in "intersexual love" and reproduce in vivo, the Professorials and other dissenters have been profoundly affected by those psychological and social forces. When the novel opens, they have given in, and "agree with the code in principle" (16). Yet as the plot begins to unfold, the code is being questioned for the first time, not only by Professorials and dissenters, but also by members of the Council themselves.

Mitchison describes herself in the jacket copy to *Solution Three* as "interested in the behavior of people who find themselves in difficult or responsible situations." The novel explores just such a situation: "solution three," the code's project of reproductive control, has begun to unravel in the face of botanical, biological, and erotic anomalies. The novel dramatizes the responses to these difficult and disturbing assaults on the code by three people possessing both responsibility and power, as members of the global Council: Jussie, who supervises the deviant Professorial plant geneticist Miryam and the clone mum Lilac; Ric, the Council historian; and Matumba, the politician who heads the Council.

A series of unforeseen events confronts the three Council members with departures from the code that challenge its exclusive commitment to the replication of identity through genetic control in both human and plant reproduction. Planning *Solution Three* in her notebooks, Mitchison summarized its central problem: "How can a good society work in an over-crowded world? (which means, central planning) How to annex the crisis of identity?"[59] The two deviant "Professorials" Miryam and Carlo, married to each other and (worse still) parents of their own biological children, discover one kind of (botanical) crisis of identity: a troubling flaw in the global crop of wheat and *rosaceae* (roses) which threatens the world's physical and emotional sustenance. Genetically engineered to be identical, the strains of wheat and roses lack the resistance that variation brings.[60] A virus appears that threatens to wipe out the entire *genus*, and the plant geneticists realize that genetic difference has acted as a biological principle aiding species survival. The biological theme has a social parallel in their own lives, for since they are heterosexuals in a society that stigmatizes that different sexual preference, they must live in cramped apartments, without the space necessary either for relaxation or productive work. Jussie, the Councillor supervising Miryam, must confront both the biological and social ramifications of this denial of difference. Miryam links her reports on the viral crisis to demands for better living space: if she is devalued because of her heterosexuality, her work as well as her personal well-being will suffer, with implications not just for Miryam but for society. Neither plants nor human beings thrive, the implication is, if variations within a population are discouraged or expunged.

Difference also appears in human relations. Lilac, one of the clone mums, refuses to surrender her clonal child to the state for "strengthening," the conditioning designed to make him develop into a replica of "Him," arguing that he should have the right to develop his own (different) identity. And a series of uprisings occur in Ulan Bator, which suggest to Council Head Matumba that the code and the "popu-policy" have not eradicated aggression. Finally, the mandated homosexuality and erotic segregation of clone children is challenged by new variations of human sexual attraction: the attraction Ric, the Council historian, feels for a clone boy who should by custom be off-limits, and the heterosexual attraction and love that two clone children feel for each other. Because the clones are siblings, not twins, their mutual attraction raises not only the specter of a stigmatized (heterosexual, incestuous) love, but more alarming still the "sin of meiosis: the upsetting of reason and planning, re-shuffling the chromosomes just anyhow!" (92) As Matumba, the head of the Council, explains: "You see, with cloning the genetic base is steady. We do not know by what chance or even not-chance He and She arrived at excellence. . . . But if two persons of different genetic excellence were to have a child the genes would be mixed. Get back to meiosis and anything can happen" (120). The mapping of religious onto genetic laws recalls Mitchison's *Saunes Bairos*, but with a twist: while in the play both kinds of law advanced patriarchal interests, here they are designed to serve interests that are not only feminist, but racially and socially emancipatory. Yet even when serving such emancipatory agendas, Mitchison suggests, an enforced religion of the genes can be oppressive.

Faced with all of these challenges to the code, and its goal of using social and genetic control of reproduction to produce a stable society, the characters in *Solution Three*, like the citizens in Vera Brittain's *Halcyon, or the Future of Monogamy* (1929), reassess the value of technologies of reproductive control and decide to loosen their grip a bit. Just as Brittain's future citizens evaluate and reject ectogenesis because of its negative social and personal consequences, so the Councillors begin to realize that the value of diversity may exceed the value of control. In response, they begin to reassess the code itself, as well as the principles of genetic and social uniformity it embodies.

Together, their responses dramatize a set of alternative scientific practices. Ric, the Council historian, comes to realize that the code, which he had understood as objective scientific data, was in fact the product of interpretation: "You see, about the code. We take it for granted but it wasn't always there. Actually, She had left notes for it, which we try to follow. But there were questions of interpretation. . . . She had even occasionally used a kind of medical shorthand. It was thought I was good at this interpretation" (78). Just as the code comes to be understood as an interpretive construct, so too Matumba begins to rethink the implications of a biological fact: meiosis. This random splitting of chromosomes, which leads to a random recombination of them in the act of fertilization, comes to seem less a "sin" to her than it has in the past.[61] Although the control made possible by the code now appears more problematic, because it introduces biological vulnerability along with social stability, so too the randomness of meiosis comes to seem more beneficial than it once did, because it introduces variability and flexibility along with unpredictability. Councillor Jussie makes a space for this beneficial vulnerability when she tells Lilac, the rebellious clone mum, that genes may *not* be all that matters: the maternal gestational environment may also be an important influence on the developing fetus. "There is . . . some evidence of non-chromosomal maternal influence from the cell material" (98). Although she acknowledges that "the watchers and carers minimize this," Jussie suggests that Lilac may want to investigate the

> "rather unexpected material about other influences on the Clones. It is not merely the physical difference between one uterus and another, but there is also considerable interchange of fluids between the foetus and its host." . . .
>
> "Also," said Jussie, "there is some difference in maternal influence

during the first two years. The watchers and carers minimize this. But it exists." (99)

Jussie's suggestion that Lilac devote scientific time to a possibility minimized by those holding greatest social power implicitly acknowledges that social and ideological agendas shape the construction of scientific facts, and embodies an alternative, more inclusive approach to the setting of scientific questions.[62] Finally, Matumba herself puts into words her increasing feeling that the value of diversity (genetic and social) may exceed the value of control: "I'm feeling my way. . . . Could be towards Solution Four. I don't know yet. But it could be that we can't afford to let too many human genes slip out. In case there was need some time. Unworthy. Of course we are. But the need might come" (154).

Near the end of *Solution Three*, there is an interchange that embodies Mitchison's feminist revision of scientific practice. Reviewing the uprisings in Ulan Bator, the crop failures, and the changes in clone sexuality, Councillor Matumba suggests that it may be time to reverse the scientific policy limiting reproduction to cloning, and thus controlling and limiting the gene pool. "But we can't just reverse a policy without landing the world in guilt and misery!" her fellow councillor Jussie protests. "That's not the way it should be done," replies Matumba. "Nor will it. None of us need be guilty. We have done right. But also we can change that right. Gently" (155).

SOLUTION THREE AND SHULAMITH FIRESTONE

If we step back a bit from *Solution Three* to consider the situation that Mitchison has invented for its characters, we can see that it functions as an intervention into two interwoven levels of discourse: contemporary biology and contemporary feminist thought. In biological terms, as I have shown, Mitchison's novel reflects her engagement with biology, particularly with the work of James D. Watson on "the master molecule," DNA. Entering the debate over the locus of embryological determinism, she weighs in (against Watson) on the side of a complex interaction between genetic material and environment, whether that is understood in terms of cellular development (influence of cytoplasm as well as nucleus), or in terms of human gestational development (importance of the gestating clone mum as well as the chromosomal heritage of the cloned child).

Although the biologically based situation Mitchison invents for her characters reflects the state of discourse within contemporary biology, her novel can also be read as a meditation on an issue introduced to feminists by a book appearing the very year Mitchison wrote her novel, and fully five years before she published it. Shulamith Firestone's 1970 *Dialectic of Sex* maintained that female emancipation required "the freeing of women from the tyranny of their reproductive biology by every means available," including "the more distant solutions based on the potentials of modern embryology, that is, artificial reproduction."[63] This "utopian" (or, many might say, dystopian) future in which artificial reproduction supersedes pregnancy did not extend to the "boundless faith in technology as a means of transcending anatomical destiny" for which one contemporary feminist critic—Margaret Talbot—has criticized Firestone's work.[64] Rather, Firestone qualified her technological optimism, granting that "in the hands of our current society and under the direction of current scientists (few of whom are female or even feminist) any attempt to use technology to 'free' anybody is suspect" (206). But Firestone did argue that "artificial reproduction" possessed the crucial potential "to threaten the *social* unit that is organized around biological reproduction," and she saw the abolition of that biologically based social unit as essential to women's liberation (206).

In the three decades since the publication of Firestone's work, her attack on pregnancy in the name of an emancipatory vision of artificial reproduction has been drowned out by a new, highly influential, radical feminist discourse on reproduction that emphasizes the oppressive, objectifying practices of reproductive technology.[65] And although *The Dialectic of Sex* was named one of the "twenty most influential 'women's' books of the past two decades" by *Publisher's Weekly*, it is currently "not only out of print but out of cultural circulation."[66]

By its very disappearance, *The Dialectic of Sex* testifies to the crucial importance for feminism, and in particular for the feminist critique of science, of the issues it raises. In her discussion of "the science question in feminism," Sandra Harding has urged "open acknowledgement . . . of certain tensions that appear in the feminist critiques [of science]," observing that "it is the tensions we long to repress, to hide, to ignore, that are the dangerous ones" (*The Science Question in Feminism*, 243). With Harding's insightful observation in mind, I suggest that what made Firestone's tract so powerful, and what subsequently also accounted for its cultural repression, was the profound, unarticulated tension in the feminist response to the scientific control project.

Feminism has vehemently deplored the project of achieving scientific control over reproduction, understanding it as part of the broader objectifying and instrumentalist project of Enlightenment science. But in *The Dialectic of Sex*, Firestone spoke the unspeakable, asserting its appeal *if it were put to feminist uses.*

Mitchison's novel explores that repressed tension within the feminist critique of science and articulates the desires and fears generated by the converging projects of scientific and social control over reproduction, emblematized by Firestone's feminist tract and Watson's theory of the "master molecule." Whereas the former has a feminist agenda, the latter demonstrates an oppressive gender ideology in its disregard for environmental and maternal influences on development. But *Solution Three* does more than represent that tension; it shows how it might be resolved, on both biological and social levels. On the biological level, Mitchison dramatizes how a broader conception of the scientific project can be used to reorient the scientific agenda (as Lilac's experience as a clone mum motivates the scientific investigation of gestational influence in embryonic development). On the social level, she represents dramatically the scientific importance of social diversity, to catalyze a reconstruction of the sites of scientific experimentation, and the social uses to which science is put, in response to changing social needs.

In dramatizing this alternative feminist vision of scientific practice, Mitchison's *Solution Three* anticipates philosopher Helen Longino's vision of "science as social knowledge." Longino describes feminist science as one that while existing in dialectical tension with mainstream science, is linked in respectful relation to a local scientific community and proceeds by the constant reincorporation of excluded scientific conclusions, measuring its findings not against some ultimate abstract standard of truth, but rather "against the cognitive needs of a genuinely democratic community" (*Science as Social Knowledge*, 214). "A consequence of embracing the social character of knowledge," Longino observes wryly, "is the abandonment of the ideals of certainty and of the permanence of knowledge. Since no epistemological theory has been able to guarantee the attainment of those ideals, this seems a minor loss" (232). Against that minor loss, Mitchison's model for a feminist science establishes the conditions for what could be a major social gain, both in scientific knowledge and in the social applications of that knowledge. As a postscript, a specific example can illustrate the concrete social implications of Mitchison's vision.

POSTSCRIPT: FROM SURROGATE
MUMS TO CLONE MUMS

The vision of feminist-enforced surrogacy in *Solution Three* seems as audaciously counterintuitive today as Firestone's praise of the emancipatory potential of "artificial reproduction" in *The Dialectic of Sex*, and it is not surprising that both books have been out of print for some time.[67] Both works articulate the initially uncomfortable, disruptive notion that feminists may not need to renounce science wholly. I am thinking here of Donna Haraway's stubborn and brave refusal to surrender the noninnocent pleasures of science: "The connections made possible by the knowledge-producing practices, and their constitutive narratives, of techno-science . . . the viscous, physical, erotic pleasure we [experience] from dis-harmonious conversations about abstract ideas, auto repair and possible worlds."[68] My point here is *not* that Naomi Mitchison persuades us to accept ectogenesis, cloning, or surrogacy, but rather that she makes us aware of more options in our responses to those, as to any, scientific technologies. She shows us that as feminists we may still be able to use, and profit from, scientific knowledge, if we engage in the feminist transformation of scientific epistemologies and practices mapped out in Mitchison's fiction, particularly in *Solution Three*.

If we compare Mitchison's tale of the clone mums to the situation of today's surrogate mothers, we can formulate an understanding of how a feminist science can reshape both what science investigates (and thus, what its findings are), and how scientific knowledge is deployed in society. Journalist Elizabeth Wasserman has explained the motivation behind an increasing tendency for states to outlaw surrogacy: "At the root of concern over surrogacy was [the fact] that the contracts legally separated the genetic aspect of reproduction from childrearing—striking at the heart of long-held ideals about the family."[69] This tendency to privilege the genetic over the gestational contribution, reflecting the fact that it is the genetic parent who is in the economically more powerful position of "commissioning" the pregnancy from the gestational parent, also characterized the 1990 case of "Baby Boy Johnson." In this case, a California judge awarded rightful motherhood not to "Anna Johnson, the black 'gestational surrogate' who, for $10,000, carried him and birthed him, but [to] Crispina Calvert, the wombless Asian-born woman who provided the egg from which, after in vitro fertilization with her (white) husband's sperm and implantation in Ms. Johnson, the baby grew."[70]

What might it mean if Naomi Mitchison's model for feminist science were put into practice in our society? It would mean a broader range of areas subject to scientific investigation and application, for the scientific agenda would reflect not only those in power, but those marginal to power. We can compare the situation of the clone mums to that of Anna Johnson, surrogate "mum" to Baby Boy Johnson. The clone mums begin by accepting a position very like the ruling under dispute in the "Baby Boy Johnson" case: "The Clones didn't belong to the Mums who had been their nests and love givers, but whose own cell nucleus had been eliminated at an early stage" (35). But because the scientists and social policy experts in their society accepted the notion that scientific practices change in response to local knowledge and the needs of the local community, the Councillors in Mitchison's society meet a clone mum's reluctance to surrender her clone boy to the state for "strengthening" not by litigating, but by listening. They use her dissident perspective as a corrective (from the margins) to centrist science and social policy in a fashion that, according to Helen Longino, exemplifies feminist scientific practices.

The clone mum Lilac's dissenting and marginal viewpoint even sets an alternative agenda for science. Councillor Jussie urges her to follow up on certain interesting (albeit not yet scientifically accredited) findings that suggest a greater role for the gestational environment and the early nurturant mother in embryonic and early infant development. As a result of this feminist reconceptualization of science, the hegemony of a genetic construction of parenthood that has been "solution three" ends, replaced by "solution four," a more diverse notion of the fetus/child as subject to genetic, gestational, and environmental influences, all of them worthy of scientific investigation and application.

Mitchison has described her work as uncomfortably positioned between discursive fields: "Scientists think I am frivolous and nonscientists think I make things difficult."[71] In the jacket copy to *Solution Three*, she describes herself as asking "awkward questions of her three scientific sons, who between them fill three Professorial chairs, and of their friends and colleagues." Yet it is by precisely that willingness to make trouble, to ask difficult questions about boundaries of science and the subjects and objects of scientific knowledge, that Naomi Mitchison has contributed powerfully to the feminist vision of an emancipatory science.

Notes

1. *New Age*, January 1986. My thanks to Rebecca Albury, of the University of Wollongong, NSW, Australia, for this image.

2. *New York Times Book Review*, 2 April 1989, 11.

3. *Sunday Mail* (Australia), 11 March 1990.

4. Robin Cook, *Mutation* (New York: G. P. Putnam & Sons, 1989), 317.

5. In the IVF procedure, the test tube holds only a fertilized ovum. In contrast, these images literalize that phrase and collapse its developmental trajectory, figuring a baby in a bottle. My understanding of the image of babies in bottles as a cultural icon, possessing both aesthetic and scientific value, is indebted to Michael Bryson's reading of the iconic function of the double helix of DNA. Bryson, "Model Transformations: The Double Helix, Gender, and Popular Literature" (Department of English, Virginia Polytechic University, photocopy).

6. The Warnock Report provides the best definition of these terms: *parthenogenesis* is "the reproductive process whereby a gamete develops into a new individual without fertilisation"; *cloning* "is the production of two or more genetically identical individuals. . . . One method of achieving cloning would be by division of the embryo at a very early stage of development so that identical genetic material is passed on to each of the separate portions." Mary Warnock, *A Question of Life: The Warnock Report on Human Fertilisation and Embryology* (Oxford: Basil Blackwell, 1985), 72. Hereafter cited by page number as the Warnock Report in both text and notes.

7. Andrew Ross, *Strange Weather: Culture, Science, and Technology in the Age of Limits* (London: Verso, 1991), 19.

8. I have, somewhat tongue in cheek, borrowed the term "situated knowledges" from Donna Haraway, who makes it a meaningful central argumentative point in her brilliant *Simians, Cyborgs, and Women: The Reinvention of Nature*

201

(New York: Routledge, 1992). I address this term seriously in Chapter 5, but here I use it simply to refer to the uninterrogated politics of *New Age*.

9. Stephen Dobyns, "Hoping for Something Worse," review of *Geek Love*, by Katherine Dunn, *New York Times Book Review*, 2 April 1989, 11. For an analysis of the novel as a work of "cyborg subjectivity," see N. Katherine Hayles, "The Life Cycle of Cyborgs: Writing the Posthuman," in *A Question of Identity: Women, Science and Literature*, ed. Marina Benjamin (New Brunswick: Rutgers University Press, 1993), 152–170.

10. Cook's epigraph is from Mary Shelley, *Frankenstein, or the Modern Prometheus* (1818).

11. Beginning in Ellen Moers's *Literary Women* with a reading of the novel as an autobiographical birth myth reflecting Mary Shelley's own troubled reproductive life, continuing with Judith Wilt's delineation of the novel's debt to the gothic genre (in its depiction of the male motive to overthrow maternal power and become the source of all life), and Gilbert and Gubar's reading of the novel in the tradition of aesthetic self-recreations by Milton's daughters, feminist readings of the novel have emphasized its preoccupations with—and critique of—the processes of conception, mothering, and being mothered. Ellen Moers, *Literary Women* (New York: Doubleday, 1974); Judith Wilt, "*Frankenstein* as Mystery Play," *The Endurance of Frankenstein: Essays on Mary Shelley's Novel*, ed. George Levine and U. C. Knoepflmacher (Berkeley and Los Angeles: The University of California Press, 1979); Sandra Gilbert and Susan Gubar, *The Madwoman in the Attic: The Woman Writer and the Nineteenth-Century Imagination* (New Haven: Yale University Press, 1979); and Anne K. Mellor, *Mary Shelley: Her Life, Her Fiction, Her Monsters* (London: Methuen, 1988).

12. Lennart Nilsson, *A Child Is Born*, rev. ed. (New York: Dell, 1977).

13. Of course, to say that the visual images do not invoke race explicitly is not to say that race plays no part in their representations. Rather, Cook portrays fetuses who *seem* white because their race is unmarked.

14. See Donna Haraway, *Primate Visions: Gender, Race, and Nature in the World of Modern Science* (London: Routledge, 1990), esp. chap. 7, for an analysis of the implications of this image of the chimpanzee/human touch.

15. I discuss *Mutation* at greater length in my essay "Conceiving Difference: Reproductive Technology and the Construction of Identity in Two Contemporary Fictions," in *A Question of Identity: Women, Science, and Literature*, ed. Marina Benjamin (New Brunswick: Rutgers University Press, 1993), 97–115.

16. Kuhn's definition of *paradigm* is elusive in *The Structure of Scientific Revolutions*. In the body of the work he describes paradigms as achievements that share at least the following two characteristics: they produce adherents, and they generate problems for the adherents to solve. But in the postscript he defines a paradigm as "what the members of a scientific community share," and goes on to say, "conversely, a scientific community consists of men [*sic*] who share a paradigm." Kuhn attempts to distinguish two basic usages for the term "paradigm": the "disciplinary matrix" of a scientific field (including its

"symbolic generalizations," its "belief in particular models," and its values), and the examples shared by a scientific field. A *paradigm shift* is a "transition between incommensurables," a "transfer of allegiance from paradigm to paradigm." For my purposes here, what is most interesting about paradigms is that they are often contained and perpetuated by the literary structures and forms that escape our examination when we study the production of scientific knowledge: by analogies, origin stories, metaphors. Thomas S. Kuhn, *The Structure of Scientific Revolutions*, 2d ed., enlarged (Chicago: University of Chicago Press, 1970), 176, 182–185, 150–151.

17. Joseph Rouse, "What Are Cultural Studies of Scientific Knowledge?" *Configurations: A Journal of Literature, Science, and Technology* 1 (winter 1993): 1.

18. Donna Jeanne Haraway, *Crystals, Fabrics, and Fields: Metaphors of Organicism in Twentieth-Century Developmental Biology* (New Haven: Yale University Press, 1976), 2.

19. As Rouse itemizes them, the most important theoretical tenets of this new field of inquiry are "antiessentialism about science; a nonexplanatory engagement with scientific practices; an emphasis upon the materiality of scientific knowledge; an even greater emphasis upon the cultural openness of scientific practice; subversion of, rather than opposition to, scientific realism or conceptions of science as 'value neutral'; and a commitment to epistemic and political criticism from within the culture of science." Rouse, "Cultural Studies," 7, 4.

20. My use of the term "literature" is not a specialized one here. I mean it to indicate fictional representations from the popular to the canonical, produced for and consumed by a range of discursive communities.

21. As Belsey observes, "Literature represents the myths and imaginary versions of real social relationships which constitute ideology." Moreover, "Literature as one of the most persuasive uses of language may have an important influence on the ways in which people grasp themselves and their relation to the real relations in which they live." Catherine Belsey, "Constructing the Subject: Deconstructing the Text," in *Feminisms*, ed. Robyn R. Warhol and Diane Price Herndl (New Brunswick: Rutgers University Press, 1991), 593, 598.

22. Bruno Latour and Steve Woolgar, *Laboratory Life: The Construction of Scientific Facts*, rev. ed. (Princeton: Princeton University Press, 1986).

23. Allan G. Gross, *The Rhetoric of Science* (Cambridge: Harvard University Press, 1990).

24. Sandra Harding, *The Science Question in Feminism* (Milton Keynes: Open University Press, 1986), 208.

25. As Harding explains: "The accounts by scientists themselves, as well as by philosophers and historians, of the famous institution that has advanced their personal and professional lives are limited by these scholars' perceptions of what is significant about the history of science; by the deficiencies of their resource materials; by unwillingness or inability to acknowledge and account for whatever compromises were made in the process of gaining social recognition and social support for their beliefs and practices; and by the inadequate conceptual schemes of the social sciences more generally, which limit our

understanding of the forces and desires that have directed social change."
Ibid., 209.

26. I have borrowed the term "symptomatic reading" from Jane Gallop, *Around 1981: Academic Feminist Literary Theory* (New York: Routledge, 1992), 7.

27. Robert Edwards and Patrick Steptoe, *A Matter of Life: The Story of a Medical Breakthrough* (London: Sphere Books, 1981), jacket copy.

28. As Naomi Mitchison explains in her memoirs, "the Huxleys were so nearly counted as kin that chaperonage was not considered necessary." Mitchison, *All Change Here*, in *As it Was: An Autobiography, 1897–1918* (Glasgow: Richard Drew Publishing, 1988), 71.

29. Carole Pateman, *The Sexual Contract* (Stanford: Stanford University Press, 1988), 95–96.

30. See note 11.

31. A fascinating parallel exists between the development of a modern notion of the "literary" as comprising that which focuses on an individual subject undergoing experiences that lead to (internal psychic) growth and the surge in modern times of interest in embryology and evolution. See Andrea Henderson, "Doll-Machines and Butcher-Shop Meat: Models of Childbirth in the Early Stages of Industrial Capitalism," *Genders* 12 (winter 1991): 100–119. See also Clifford Siskin, *The Historicity of Romantic Discourse* (New York: Oxford University Press, 1988), esp. chap. 5.

32. Henderson suggests that the anxiety aroused by such mechanistic constructions of the woman in childbirth may have led William Hunter to offer a compromise in his obstetrical atlas: he would no longer figure the woman as a machine, but rather would show her as a fleshly, but not definably human, animal. Henderson, "Doll-Machines," 100.

33. "Men appropriate to themselves women's natural creativity, their capacity physically to give birth—but they also do more than that. Men's generative power extends into another realm; they transmute what they have appropriated into another form of generation, the ability to create new political life, or to give birth to political right." Pateman, *Sexual Contract*, 88.

34. See Bernard Doray, *From Taylorism to Fordism: A Rational Madness*, trans. David Macey (London: Free Association Press, 1988); Edward Yoxen, *The Gene Business: Who Should Control Biotechnology* (London: Free Association Books, 1986); and Emily Martin, *The Woman in the Body: A Cultural Analysis of Reproduction* (Boston: Beacon Press, 1987), esp. chap. 2.

35. Daniel J. Kevles, *In the Name of Eugenics: Genetics and the Uses of Human Heredity* (Berkeley and Los Angeles: University of California Press, 1985), 173.

36. Gary Werskey, *The Visible College: A Collective Biography of British Scientists and Socialists of the 1930s* (London: Free Association Books, 1988).

37. Naomi Mitchison, *You May Well Ask: A Memoir, 1920–1940* (London: Flamingo Books, 1986), 34–35.

38. Naomi Mitchison, ed., *An Outline for Boys and Girls and Their Parents* (London: Victor Gollancz, 1932), 419.

39. Editorial, *The Realist: A Journal of Scientific Humanism* 1 (April-June 1929): 179.

40. Naomi Mitchison, *Comments on Birth Control* (London: Faber & Faber, 1930).

41. Norman Haire, ed., *Sexual Reform Congress* (London: Kegan Paul, Trench, Trubner, 1930), xxii–xl.

42. As the advertisement for the series explains, the contributions were "written from various points of view, one book frequently opposing the argument of another." Advertisement for the To-Day and To-Morrow series, back pages following J. D. Bernal, *The World, the Flesh, and the Devil: An Enquiry into the Future of the Three Enemies of the Rational Soul* (London: Kegan Paul, Trench, Trubner, 1930), 97.

43. See especially Joanna Russ, "When It Changed," in *The Zanzibar Cat* (Sauk City, Wis.: Arkham House, 1983), 3–11; idem, "The Clichés from Outer Space," in *The Hidden Side of the Moon* (New York: St. Martin's Press, 1987); idem, *The Female Man* (Boston: Beacon Press, 1975); Marge Piercy, *Woman on the Edge of Time* (New York: Fawcett Crest, 1976); Margaret Atwood, *The Handmaid's Tale* (New York: Fawcett Crest, 1985); Octavia Butler, *Dawn* (New York: Warner Communications, 1987); idem, *Adulthood Rites* (New York: Warner Books, 1988); idem, *Imago* (New York: VGSF, 199); Elizabeth Jolley, *The Sugar Mother* (New York: Harper & Row, 1988); Angela Carter, *The Passion of New Eve* (London: Virago, 1977); and Fay Weldon, *The Cloning of Joanna May* (London: Collins, 1989).

44. Weldon's *Cloning of Joanna May* is most explicitly emancipatory in its representation of reproductive technology, although that representation is the result of considerable work within the plot to refigure a technology that is, initially, oppressive. Joanna Russ, Elizabeth Jolley, and Octavia Butler also tend toward emancipatory interpretations, whereas Marge Piercy suggests the technology's potential for good or ill, and Margaret Atwood renders a terrifying portrait of its capacities for objectification and control. Of literary critics, Donna Haraway's celebrated praise of cyborgs exemplifies the positive literary critical interpretation, whereas Gena Corea's work most exemplifies the negative feminist theoretical response to RT. Donna Haraway, "A Manifesto for Cyborgs: Science, Technology, and Socialist Feminism in the 1980s," *Australian Feminist Studies* 4 (autumn 1987): 1–42, and Gena Corea, *The Mother Machine: Reproductive Technologies from Artificial Insemination to Artificial Wombs* (London: The Women's Press, 1988).

45. Michelle Stanworth, "Birth Pangs: Conceptive Technologies and the Threat to Motherhood," in *Conflicts in Feminism*, ed. Marianne Hirsch and Evelyn Fox Keller (New York: Routledge, 1990), 288–304, especially 290–291.

46. N. Katherine Hayles, *Chaos Bound: Orderly Disorder in Contemporary Literature and Science* (Ithaca: Cornell University Press, 1990), 272.

47. Warnock Report, vi.

48. Colin Brown, "MPs back continued embryo research," *Independent* (London), 24 April 1990, 1.

49. "Quite apart from the parliamentary discussions themselves, the issues that led to the formulation of the Act were assumed to be in the public mind. Indeed, the earlier Committee of Inquiry led by Mary Warnock . . . had been called to respond to, if not actually allay, public concern. For many people in Britain, the concern was with the kind of future that might follow increasingly sophisticated techniques of intervention in the reproduction of human life." Marilyn Strathern, *Reproducing the Future: Anthropology, Kinship, and the New Reproductive Technologies* (London: Routledge, 1992), 4.

50. Latour and Woolgar, *Laboratory Life*, 29.

1   BABIES IN BOTTLES AND TISSUE-CULTURE KINGS

1. "Perhaps I can make the significance of this a little more clear by giving your Lordship an analogy. Exactly 10 years ago a mathematician called Mandelbrot first discovered what is now called the Mandelbrot set. It is a set of points which can be mapped out as a computer graphic to form the most amazing, beautiful and complex structure that it is possible to imagine. It is a picture of literally infinite depth. If one magnifies the details of any part of the picture, one finds that in them are whole worlds of further detail which are always beautiful, which never repeat themselves and which always reveal more and more detail, on and on, *ad infinitum.*

"How is the Mandelbrot set made? It is made by the use of an absurdly simple equation with only three terms. The secret lies in the process. It is a process whereby the answer to one use of the equation becomes the starting point for the next. In other words, it is a cumulative process, just like evolution in which one life form builds on another and just like embryology in which the development of one cell provides the context for the development of its neighbors and its successors." House of Lords, Official Record, 7 December 1989, col. 1020; quoted in Marilyn Strathern, *Reproducing the Future*, 144–145 (see intro., n. 49).

2. Ivars Peterson, "Bordering on Infinity," *Science News*, 23 November 1991, 331.

3. Evelyn Fox Keller observes that physics and physicists played a similar authority-conferring role for molecular biology: "That authority was, of course, acquired in the first place through the formidable displays of technological and instrumental power issuing from physics itself, but this initially technical authority soon became available for deployment far beyond the domain of their technical triumphs; it became, in short, an authority that could be called upon for the essentially social process of reframing the character and goals of biological science. This borrowing proceeded in a variety of ways—first, through the borrowing of an agenda that was seen as looking like the agenda of physics; second, by borrowing the language and attitude of physicists; and finally, by borrowing the very names of physicists." Chaos theory—itself a branch of physics—here authorizes a similar act of reframing in the field of human embryology. Keller, *Secrets of Life, Secrets of Death: Essays on Language, Gender, and Science* (New York: Routledge, 1992), 97–98.

4. Allan G. Gross, *The Rhetoric of Science*, 31 (see intro., n. 23).

5. Donna Jeanne Haraway, *Crystals, Fabrics, and Fields,* 10 (see intro., n. 19). Haraway is referring to Mary Hesse, *Models and Analogies in Science* (South Bend, Ind.: University of Notre Dame Press, 1966).

6. Bruno Latour and Steve Woolgar, *Laboratory Life: The Construction of Scientific Facts* (Princeton: Princeton University Press, 1986), title page.

7. Julian Huxley, "Science, Natural and Social," in *Man in the Modern World* (New York: Mentor, 1927), 127.

8. Julian Huxley, "The Tissue-Culture King," *Yale Review* 15 (April 1926): 479–504; reprinted in *Amazing Stories,* August 1927, 451–459. All subsequent references are to the earlier edition.

9. Carl Wood and Ann Westmore, *Test-Tube Conception* (London: George Allen & Unwin, 1984), 59–75, 12–13.

10. Charlotte Perkins Gilman, *Herland* (*The Forerunner,* 1915; reprinted New York: Pantheon, 1978). Fay Weldon's fictional description of the technique is a good model for what Huxley might have had in mind when he predicted its possible extension to human beings: "While she was opened up we took away a nice ripe egg; whisked it down to the lab: shook it up and irritated it in amniotic fluid till the nucleus split, and split again, and then there were four. . . . We kept the embryos in culture for four whole weeks, had four nice healthy waiting wombs at hand and on tap, for implantation. All four took like a dream: there they grew until they popped into the world, alive and kicking and well." Weldon, *The Cloning of Joanna May,* 34 (see intro., n. 44). Other parthenogenetic fictions include Joanna Russ, "When It Changed" and *The Female Man,* and Marge Piercy, *Woman on the Edge of Time* (all of which are cited in full at intro., n. 43). The nonfiction urtext for feminist discussions of parthenogenesis is, of course, Shulamith Firestone's *Dialectic of Sex* (New York: William Morrow, 1970).

11. Julian Huxley, *Essays in Popular Science* (London: Penguin, 1926), 191.

12. Julian Huxley, *Memories* (London: Penguin, 1970), 18.

13. T. H. Huxley's letter to his grandson continued: "My friend who wrote the story of the Water Baby, was a very kind man and very clever. Perhaps he thought I could see as much in the water as he did— There are some people who see a great deal and some who see very little in the same things.

"When you grow up I daresay you will be one of the great-deal seers and see things more wonderful than Water Babies where other folks can see nothing." Ibid., 20–21.

14. Charles Kingsley, *The Water-Babies: A Fairy-Tale for a Land-Baby* (1863) (London: J. M. Dent & Sons, 1957), 202. The illustrations were reproduced from the 1863 edition.

15. Gillian Beer, *Darwin's Plots: Evolutionary Narrative in Darwin, George Eliot, and Nineteenth-Century Fiction* (London: Ark Paperbacks, 1985), 130.

16. Of course, Tom dies before turning into a water-baby. Thus, one could argue that the plot figures not so much life-before-birth as reincarnation after death: Tom's transformation into a water-baby returns him to the cycle of embryonic (and moral) development. Gillian Beer also sees the fetal nature of Kingsley's water-babies, but she attributes it more to the impending influence

of Freudian psychoanalytic theory than to that of experimental zoology, arguing that "[the] book as a whole has an oceanic richness typical of just pre-Freudian storytelling, in which all the elements of primal experience are present without interpretation." Ibid., 135.

17. Although Tom expects "to find her snipping, piecing, fitting, stitching, cobbling, basting, filing, planing, hammering, turning, polishing, moulding, measuring, chiselling, clipping, and so forth, as men do when they go to work to make anything," Mother Carey doesn't need to lift a finger in order to make "old beasts into new all the year round" Kingsley, *Water-Babies*, 200, 199.

18. Julian Huxley, "The Tadpole," in *Essays in Popular Science*, 232.

19. Huxley, *Memories*, 209.

20. Ibid., 119; J. F. Gudernatsch, *American Journal of Anatomy* 15, no. 4 (1914): 431.

21. Juliette Huxley, *Leaves of the Tulip Tree* (London: John Murray, 1986), 106.

22. Huxley, *Memories*, 120. *Daily Mail*, 20 February 1920. Follow-up reports announced, "This is one of the hints of the power that man can yield by the chemical treatment of the egg or embryo," and presented Huxley himself as a scientific case study, proclaiming his "heredity of genius." *Daily Mail*, 23 February 1920, 7, and 24 February 1920, 7.

23. "Young Huxley," *Daily Mail*, 20 February 1920, 1.

24. Julian Huxley, "Secrets of Life. Mr. Huxley on His Clues," *Daily Mail* 25 February 1920, 9. See also Huxley, *Memories*, 120.

25. For a contemporary discussion of such uses of hormone therapy to police behavior, see Jennifer Terry, "Stories Animals Tell: Predicaments of Gender in Recent Biological Research on Sexual Orientation" (paper presented at the Stony Brook Humanities Institute, Stony Brook, New York, 15 September 1992).

26. In his 1922 essay "Sex Biology and Sex Psychology" Huxley assesses the relative authority of biological and psychoanalytic accounts of sex difference, abnormal sexual behavior, and sexual difficulties. The conclusion decisively favors biology, which Huxley argues explains most of the "abnormalities of sexual psychology" as the result of "simple physical abnormalities." Education in developmental and evolutionary biology can provide the informational grounding essential to a successful sexual life as an adult, he further argues, without the paralyzing self-consciousness produced (in his view) by psychoanalytic treatment. Julian Huxley, "Sex Biology and Sex Psychology," *Essays of a Biologist* (New York: Alfred A. Knopf, 1923), 132–173, 171. Huxley's position against psychoanalysis reflects not just his era, but his own experiences. Huxley had a nervous breakdown in 1913 which he linked to the complications of his affair with "K," a woman toward whom he felt "attraction, loyalty, and guilt," but with whom he was not "in love . . . in the true sense of the word." Another nervous breakdown followed his honeymoon in 1919, and although Huxley attributes it to his difficulties in a new teaching job, another explanation is suggested by his description of the honeymoon

itself: "Juliette always maintains that I spent the week absorbed in the love-making of grebes, while she had to wait outside the hide in the bitter April winds, without any practical study of human love-making. I am afraid there is some truth in this: I was too keen to get on with my scientific studies of bird display." Huxley, *Memories*, 92, 117.

27. See Norman Haire, "Rejuvenation," *The Realist: A Journal of Scientific Humanism* 1 (April 1929): 38–47; and Huxley, "Sex Biology and Sex Psychology," 139. The focus of scientific and medical interest through the next decade, rejuvenation experiments would also figure widely in popular science essays as well as fiction, including works by Aldous Huxley, most notably *Ape and Essence* (London: Triad Grafton, 1985). In July 1939, Charlotte Haldane's magazine, *Woman Today*, published an article on endocrine treatments by Dr. Barbara Holmes, M.A., Ph.D., "The Gland that Controls your Sex."

28. Julian S. Huxley, "Searching for the Elixir of Life," *Century Illustrated Monthly Magazine*, February 1922, 626.

29. These issues are all discussed at length in the debate over the Human Fertilisation and Embryology Bill. Consider the striking similarity between Huxley's prophetic portrait of the body's *regeneration*, and Barbara Katz Rothman's more recent description of the body, in a study of human *generation*: "The Cartesian model of the body as a machine operates to make the physician a technician, or mechanic. The body breaks down and needs repair; it can be repaired in the hospital as a car is in the shop; once 'fixed,' a person can be returned to the community. . . . Capitalism adds that not only is the body a collection of parts; its parts become commodities. In the United States, the essential fluids of life—blood, milk, and semen—are all for sale." Rothman, *In Labor: Women and Power in the Birthplace* (New York: W. W. Norton, 1991), 34–35.

30. "Thyroid Feeding and Growth," *Lancet*, 28 April 1923, 861.

31. Huxley, "The Tadpole," 192–193. Of course, the implications of this field resonated back into history, and forward to the present day. In Huxleys version, experimental embryology originated in the field of descriptive embryology founded between 1780 and 1820 by Wolff, Goethe, Pander, and von Baer, whose primary insight was that "development was . . . truly epigenetic [producing an 'entirely new structure during embryonic development'] and no mere unfolding of a preformed miniature" (192). Thus the field represents two decisive breaks in the conceptualization of reproduction: the abandonment of the homunculus theory and the contemporary movement into research on human embryos. In its focus on development and mechanisms of differentiation, experimental embryology anticipates the questions and issues central to the debate over the Human Fertilisation and Embryology Bill: When does a new identity come into being in the developmental process, and how does the process of differentiation help in its production? Huxley, "The Tadpole"; M. Abercrombie, C. J. Hickman, and M. L. Johnson, *The Penguin Dictionary of Biology*, 7th ed. (London: Penguin Books, 1980), 106; Strathern, *Reproducing the Future*.

32. Natalie Angier, "Making an Embryo: Biologists Find Keys to Body Plan," *New York Times*, 23 February 1993, C1. The Hox genes are the developmental genes found to enfold the so-called homeobox sequence, the molecular sequence that scientists believe triggers developmental unfolding. As Natalie Angier explains, "The homeobox is the so-called binding domain of the gene. When a Hox gene is translated into a protein to perform its chore within the cell, the homeobox specifies a little corkscrew shape unit that allows the entire Hox protein to behave as a transcription factor, twisting around the double helix in a sinuous hug and then switching on the genes" (C9). Adrienne Munich, in her forthcoming book on Queen Victoria, suggests that the proliferation of images of Queen Victoria in the latter years of her reign arose from an anxiety about the conjunction of state power and the female body. The use of an image of the queen to embody the principle of developmental differentiation is a particularly concrete and pointed instance of such a phenomenon. See also Adrienne Munich, "Queen Victoria, Empire, and Excess." *Tulsa Studies in Women's Literature* 6 (fall 1987): 265–281.

33. A contemporary definition does not stray from the meaning it had in Huxley's time: "artificial activation of development of an egg . . . without contact with sperm." Abercrombie, Hickman, and Johnson, *Penguin Dictionary of Biology*, 29.

34. Julian S. Huxley, "The Determination of Sex," in *Essays in Popular Science*, 43.

35. Bill 106, also known as the Human Fertilisation and Embryology Bill, debated and passed by Parliament in 1990, defines the "father" of a child produced by reproductive technology in terms not of his genetic or nurturant relation to the child, but his relationship to the scientific intervention producing the pregnancy. Thus consent to the process of reproductive technology constitutes a man as a father. The bill specifies that if a woman undergoing IVF or *AID* (artificial insemination by donor) is married at the time of the procedure, "the other party to the marriage shall be treated as the father of the child unless it is shown that he did not consent to the placing in her of the embryo or the sperm and eggs or to her insemination (as the case may be)." If the woman is not married, but "(a) the embryo or the sperm and eggs were placed in the woman, or she was artificially inseminated, in the course of treatment services provided for her and a man together by a person to whom a license applies, and (b) the creation of the embryo carried by her was not brought about with the sperm of that man, then . . . that man shall be treated as the father of the child." *Human Fertilisation and Embryology Bill* (H.L.) (Bill 106) (London: HMSO, 21 March 1990), 15.

36. Abercrombie, Hickman, and Johnson, *Penguin Dictionary of Biology*, 298.

37. The popular press was also responsible for Carrell's fame. When he visited London in 1924, the *Daily Express* ran a story on him with four layers of shrill headlines: "ALIVE WITHOUT A BODY. HEART THAT THROBS BY ITSELF. TWELVE YEARS. US WONDER SURGEON HERE. *Daily Express*, 21 July 1924, 3. If endocrine treatment exploited associations reaching back to myth, tissue culture

drew on associations reaching forward into scientific rationality. As Andrew Ross observes: "In the fledgling struggles over genre formation in the SF of the twenties and thirties, what we can see is a contest to establish a language that signified scientific rationality and to eliminate a language that privileged romance, fantasy, and literary invention—except, of course, where the romance was that of science and technology." Ross, *Strange Weather*, 113 (see intro., n. 7). There are gendered implications to this elimination of romance, fantasy, and invention, as I discuss in relation to Charlotte Haldane's treatment of science in Chapter 3.

38. Editor's introduction to A. Hyatt Verrill, "The Plague of the Living Dead," *Amazing Stories*, April 1927, 7. This short story, most likely the original version of the cult film *Night of the Living Dead*, interweaves the familiar Huxleyan themes of life-extension therapy, fetal gestation, and rejuvenation.

39. A. Hyatt Verrill, "The Ultra-Elixir of Youth," *Amazing Stories* August 1927, 476; Francis Flagg [George Henry Weiss], "The Machine Man of Ardathia," *Amazing Stories* November 1927, 798–804.

40. James Gunn, ed., *The New Encyclopedia of Science Fiction* (New York: Viking, 1988), 174.

41. Cf. the illustration in Wood and Westmore, *Test-Tube Conception*, 62.

42. This may be an allusion to the earlier reference to Carrel's research, in Verrill, "The Plague of the Living Dead," 7.

43. Gunn, *New Encyclopedia of Science Fiction*, 16.

44. Ibid. Gernsback is a central figure in Andrew Ross's historical analysis of the contest between pro-scientific boosterism and "critical technocracy" in the shaping of pulp SF. Ross, *Strange Weather*, 105.

45. E. M. Forster, "The Novels of Virginia Woolf," *Yale Review* 15 (April 1926): 505.

46. "I suddenly noted something . . . which made me feel queer—a telephone wire, with perfectly good insulators, running across from tree to tree. A telephone—in an unknown African town. I gave it up." Huxley, "The Tissue-Culture King," 482. In the later editions of this story, the passage was changed to emphasize not the narrator's emotions, but his rationality: "I suddenly noted something else which appeared inexplicable—a telephone wire, with perfectly good insulators, running across from tree to tree. A telephone—in an unknown African town." Huxley, "The Tissue-Culture King," *Amazing Stories*, 452; reprinted in Groff Conklin, ed., *The Best of Science Fiction* (New York: Crown Publishers, 1946), 348–365.

47. As Marilyn Strathern has observed, "The faculty of making analogies . . . is central to the cultural life anywhere." Strathern, *Reproducing the Future*, 2.

48. Gena Corea, *The Mother Machine*, 2 (see intro., n. 44).

49. Londa Schiebinger, *The Mind Has No Sex? Women in the Origins of Modern Science* (Cambridge: Harvard University Press, 1989).

50. For discussions of the role of animal experimentation in the development of reproductive technology in human beings, see Edward Yoxen, "Historical Perspectives on Human Embryo Research," in *Experiments on*

Embryos, ed. Anthony Dyson and John Harris (London: Routledge, 1990), 27–41, and Corea, *The Mother Machine.*

51. The fictional laboratory devoted to endocrine experimentation recalls such actual scientific studies as Huxley's axolotl experiment and the rejuvenation attempts of Steinach and Voronoff in the 1920s. As Norman Haire reported in the *Realist,* on whose editorial board Julian Huxley served, Voronoff and Steinach transplanted the glands of monkeys to men in an attempt to "remove some of the ravages of age and postpone the oncoming of senility." Haire, "Rejuvenation," 40.

52. "You know the craze for 'glands' that was going on at home years ago, and its results, in the shape of pluriglandular preparations, a new genre of patent medicines, and a popular literature that threatened to outdo the Freudians, and explain human beings entirely on the basis of glandular make-up, without reference to the mind at all." Huxley, "The Tissue Culture King," 493.

53. Gordon Rattray Taylor, *The Science of Life* (London: Thames & Hudson, 1963), 257.

54. Robert L. Duffus, "Jacques Loeb: Mechanist," *Century Illustrated Monthly Magazine,* July 1924, 383.

55. Huxley, "The Tissue-Culture King," 493. Yet in the case of the Human Genome Initiative, that control is incomplete and flawed, as Evelyn Fox Keller has pointed out: "I think that the goal of that kind of control, that kind of domination, that kind of direct 'getting your hand on the future of evolution' does lead to the selection of theories that are partial and suited to that goal. . . . The notion of control or domination that's at work is a fantasy. The world is not available to domination—I mean long-term domination." Observing that "different kinds of science would aim at different notions of control," she adds, "the notion of control that I think is built into this project [the Human Genome Project] is akin to domination." Larry Casalino, "Decoding the Human Genome Project: An Interview with Evelyn Fox Keller," *Socialist Review* 91, no. 2 (1992): 118, 124.

56. Peter Firchow, "Science and Conscience in Huxley's *Brave New World,*" *Contemporary Literature* 16 (summer 1975): 301–316.

57. This pioneering work in IVF would culminate in 1978 in the birth of Louise Brown, the first "test-tube" baby. For an account of the development of the reproductive technology that resulted in Brown's birth, see Robert Edwards and Patrick Steptoe, *A Matter of Life* (see intro., n. 27); Loretta McLaughlin, *The Pill, John Rock, and the Church: The Biography of a Revolution* (Boston, 1982); see also Corea, *The Mother Machine,* 136 n.3.

58. H. G. Wells, *The Island of Dr. Moreau* (London, 1896).

59. Lesley A. Hall, "Illustrations from the Wellcome Institute Library," *Medical History,* no. 34 (1990): 327–333. Daniel J. Kevles, *In the Name of Eugenics,* 70–84 (see intro., n. 35).

60. Donna Haraway, *Primate Visions,* 199–201 (see intro., n. 14).

61. "Eugenics, Dean Inge writes . . . is capable of becoming the most sacred ideal of the human race, as a race; one of the supreme religious duties. *In*

this I entirely agree with him. Once the full implications of evolutionary biology are grasped, eugenics will inevitably become part of the religion of the future." Julian Huxley, "Eugenics and Society," in *Man Stands Alone* (New York: Harper & Bros., 1941), 43.

Compare this wholehearted agreement with Dean Inge with the comments of Huxley's friend and fellow popular-science writer J.B.S. Haldane, who observed: "The Dean would like to penalize the slum-dwellers who still produce large families. . . . If a difference in effective fertility exists between the rich and the poor, it seems to me to be profoundly illogical to attempt to remedy it by making the rich richer and the poor poorer. . . . It would be better to send armored cars through the slums from time to time, with special instructions to fire upon women and children." Haldane, *Science and Ethics* (London: Watts, 1928), 25–26.

62. Huxley, "Eugenics and Society," 63.

63. Thomas Laqueur, *Making Sex: Body and Gender from the Greeks to Freud* (Cambridge: Harvard University Press, 1990), 161.

64. John Kobler, *The Reluctant Surgeon: A Biography of John Hunter* (New York: Doubleday, 1960), 283. As Kobler explains, Hunter's account of this operation, "posthumously published in 1799, furnished the authority for numerous uterine injections performed during the next fifty years. . . . Not until the investigations in 1866 of the American gynecologist, James Marion Sims, did artificial insemination begin to emerge as a recognized technique." *Reluctant Surgeon*, 332.

65. Herbert Brewer, "Eutelegenesis," *Eugenics Review* 27, no. 2 (1935): 121–126. Hermann Muller first saw Brewer's essay after his own book, *Out of the Night*, had already gone to press and mentions it favorably in his preface in remarks added immediately before publication (Muller, 9). For additional discussions of the history and present applications of AID, see Corea, *The Mother Machine*; Rita Arditti, Renate Duelli Klein, and Shelly Minden, *Test-Tube Women: What Future for Motherhood?* (London: Pandora Press, 1984); Michelle Stanworth, *Reproductive Technologies: Gender, Motherhood, and Medicine* (Minneapolis: University of Minnesota Press, 1987); Patricia Spallone, *Beyond Conception: The New Politics of Reproduction* (London: Macmillan, 1989); Linda Birke, Susan Himmelweit, and Gail Vines, *Tomorrow's Child: Reproductive Technologies in the '90s* (London: Virago, 1990); and U.S. Congress, Office of Technology Assessment, *Infertility: Medical and Social Choices* (Washington, D.C.: U.S. Government Printing Office, May 1988).

66. Joseph Nathan Kane, *Famous First Facts: A Record of First Happenings, Discoveries, and Inventions in American History*, 4th ed. (New York: H. W. Wilson, 1981), 319. See also Kevles, *In the Name of Eugenics*, 189.

67. Elof Axel Carlson, *Genes, Radiation, and Society: The Life and Work of H. J. Muller* (Ithaca: Cornell University Press, 1981), 393.

68. Hermann Muller, *Out of the Night: A Biologist's View of the Future* (New York, 1935; reprinted London: Victor Gollancz, 1936), 7. Muller was doing the final revisions to this book when Aldous Huxley's *Brave New World* was published. As his biographer reports, although the grim view of a eugenically

controlled world disturbed him, Muller "liked the way technology could be applied to biology and human reproduction and he claimed *Brave New World* was important in popularizing this aspect of science." Carlson, *Genes, Radiation, and Society*, 187.

69. Michel Foucault used the term "bio-power" to indicate both the systematic disciplining, regulating, and normalizing of the individual body, and also the monitoring, regulating, and control of the "species body." As he observes, "The disciplines of the body and the regulations of the population constituted the two poles around which the organization of power over life was deployed." Foucault, *The History of Sexuality*, vol. 1, *An Introduction* (New York: Vintage Books, 1980), 139.

70. Evelyn Fox Keller, "Fractured Images of Science, Language, and Power: A Post-Modern Optic, or Just Bad Eyesight?" in *Secrets of Life, Secrets of Death*, 96.

71. Donna Haraway, "A Cyborg Manifesto," in *Simians, Cyborgs, and Women*, 162 (see intro., n. 8). That exploration is not inevitably an oppressive analogy for scientific practice is dramatized by Naomi Mitchison's *Memoirs of a Spacewoman* (London: Women's Press, 1985), which figures both science and space exploration as centrally concerned not with conquest but with communication. I discuss this further in Chapter 5.

72. Evelyn Fox Keller, "From Secrets of Life to Secrets of Death," in *Body/Politics: Women and the Discourses of Science*, ed. Mary Jacobus, Evelyn Fox Keller, and Sally Shuttleworth (New York: Routledge, 1990), 177–191; "Making Gender Visible in the Pursuit of Nature's Secrets," in *Feminist Studies/Critical Studies*, ed. Teresa de Lauretis (Bloomington: Indiana University Press, 1987), 67–77; and *Secrets of Life, Secrets of Death*.

73. See Gary Werskey, *The Visible College* (see intro., n. 36), and Kevles, *In the Name of Eugenics*, for rich discussions of this conflict in the work of Huxley, Muller, and their colleagues, especially J.B.S. Haldane, J. D. Bernal, and Lancelot Hogben.

74. Huxley, "Science, Natural and Social," 127.

2   THE ECTOGENESIS DEBATE AND THE CYBORG

1. As the Warnock Report describes its terms of reference: ". . . to consider recent and potential developments in medicine and science related to human fertilisation and embryology; to consider what policies and safeguards should be applied, including consideration of the social, ethical and legal implications of these developments; and to make recommendations." Warnock Report, 4 (see intro., n. 6).

2. "Most ministers appeared to vote in favour of the crucial Clause 11 in the Human Fertilisation and Embryology Bill, based on the 1984 Warnock report. It will allow experiments on human embryos up to 14 days old under strict conditions laid down by a statutory authority." "MPs back continued embryo research," *Independent*, 24 April 1990, 1. See also the "MPs back embryo research," *Guardian*, 24 April 1990, 1.

3. Some feminist opponents to reproductive technology argue that AID,

IVF, and surrogacy are not infertility treatments, per se, since the infertile person's status does not change when by those means a baby is conceived, brought to term, and delivered. Marilyn Strathern's *Reproducing the Future* is cast as an explicit response to the Human Fertilisation and Embryology Act, as is my own article "The Human Fertilisation and Embryology Bill: Feminist Interventions and Discursive Constraints." Linda Birke, Susan Himmelweit, and Gail Vines's *Tomorrow's Child: Reproductive Technology in the '90s* (see Chap. 1, n. 65) concludes its explicit discussion of British government legislation on "infertility treatments" with the Warnock Committee report, but the implicit context for their book is the Human Fertilisation and Embryology Bill, which was still being debated in Parliament in the year *Tomorrow's Child* was published. For additional discussions of the implication of the bill, see Anthony Dyson and John Harris, *Experiments on Embryos* (London: Routledge, 1990).

4. Julian S. Murphy, "Is Pregnancy Necessary? Feminist Concerns About Ectogenesis," *Hypatia* 4 (fall 1989): 68.

5. Anne McLaren, "Reproductive Options: Present and Future," in *Reproduction in Mammals*, book 5, *Manipulating Reproduction*, ed. C. R. Austin and R. V. Short (Cambridge: Cambridge University Press, 1980), 190; cited in Birke, Himmelweit, and Vines, *Tomorrow's Child*, 209.

6. Gena Corea argues that "piece by piece, in various laboratories around the world, men are working out the technology for complete ectogenesis," and her conclusion is echoed by Peter Singer and Deane Wells, who predict that researchers will "stumble on ectogenesis" while attempting to perfect techniques for the laboratory culture of embryos and the medical treatment of immature fetuses. Gena Corea, *The Mother Machine*, 258 (see intro., n. 44); Peter Singer and Deane Wells, *Making Babies: The New Science and Ethics of Conception* (New York: Charles Scribners' Sons, 1985), 118. In contrast, Birke, Himmelweit, and Vine argue that "ectogenesis . . . is unlikely both technologically, and in economic terms, in the foreseeable future." *Tomorrow's Child*, 210.

7. Only Peter Singer and Deane Wells support the development of ectogenesis, while arguing against the in vitro gestation of nonsentient embryos as a source of spare organs. They take the utilitarian position that the technology would be welcome because it would enable us to save the lives of premature infants, render abortion unnecessary, and provide a welcome alternative to surrogacy for women unable to gestate their own children. *Making Babies*, 118–131.

8. Sigmund Freud, "Negation (1925)," *Collected Papers*, vol. 5, *Miscellaneous Papers, 1888–1938*, ed. James Strachey (New York: Basic Books, 1959), 181–185, 182.

9. J.B.S. Haldane, *Daedalus, or Science and the Future* (London: Kegan Paul, Trench, Trubner, 1923).

10. The phrases are drawn from the advertisement for the To-Day and To-Morrow series appended to Anthony M. Ludovici's *Lysistrata, or Woman's Future and Future Woman* (London: Kegan Paul, Trench, Trubner, 1927), 1. For the categories of knowledge, see the advertising matter at the back of J. D.

Bernal's book *The World, the Flesh, and the Devil: An Enquiry into the Future of the Three Enemies of the Rational Soul* (London: Kegan Paul, Trench, Trubner, 1929). The contemporary firm, Routledge, has continued the tradition of interest in the cultural studies of (reproductive) science.

11. Although he did not contribute to the To-Day and To-Morrow series, Julian Huxley himself joined the ectogenesis debate somewhat tardily, in a footnote to his memoirs. The attraction Huxley feels to such reproductive technologies is unmistakable: "The first steps toward this [ectogenesis] have now been taken, both in lower mammals and man, and early embryos have been made by fertilizing ova with sperm in a test-tube. But it is unlikely that they can ever reach full term, and many abnormalities occur as they develop. Meanwhile the outcry against 'test-tube' babies and interfering with God's laws continues, though the method opens up immense eugenic possibilities." Julian Huxley, *Memories*, 131 n. 1 (see Chap. 1, n. 12). Eight years after the publication of Huxley's memoirs, the first "test-tube baby," Louise Brown, was born in Oldham, England.

12. Mary Poovey, "Scenes of an Indelicate Character," in *The Making of the Modern Body: Sexuality and Society in the Nineteenth Century*, ed. Catherine Gallagher and Thomas Laqueur (Berkeley and Los Angeles: University of California Press, 1987), 152.

13. All of these works were published in London by Kegan Paul, Trench, Trubner, and Company. Three additional works dealt with the social implications of scientific change, but they did not explicitly comment on Haldane's notion of ectogenesis. Bertrand Russell's *Icarus, or the Future of Science* (London: Kegan Paul, Trench, Trubner, 1924) shifts the terrain under contestation away from biology, preferring to debate Haldane on the more comfortable ground of the physical and anthropological sciences. Nonetheless, to Haldane's prediction that reproductive control and management will one day be administered and monitored by committees of scientist-bureaucrats, Russell offers a prophetic analysis of the weakness of any future state ruled by "official science": "Technical scientific knowledge does not make men sensible in their aims, and administrators in the future will be presumably no less stupid and no less prejudiced than they are at present." *Icarus*, 55. C. P. Blacker's *Birth Control and the State: A Plea and a Forecast* (London: Kegan Paul, Trench, Trubner, 1926) also participates in the debate about the scientific and social control of reproduction, without giving specific attention to the implications of ectogenesis. Blacker's pamphlet, which had originally been published in the *Saturday Review*, had a specific policy agenda rather than the wider futurological scope of the standard title in the To-Day and To-Morrow series. A psychiatrist who trained in evolutionary biology under Julian Huxley, and who later became general secretary of the Eugenics Society (1932–1953), Blacker was explicitly concerned to refute the religious, nationalistic, and economic arguments against the distribution of birth-control information by medical practitioners. Daniel J. Kevles, *In the Name of Eugenics*, 171–172 (see intro., n. 35). Vernon Lee's *Proteus, or the Future of Intelligence* (London: Kegan Paul, Trench, Trubner, 1925) pillories the very con-

cept of the series, marveling at how "nowadays encyclopaedic science and journalistic emphasis are being applied to making our flesh creep with prophecies of Perils." *Proteus*, 57–58. Finally, Dora Russell's *Hypatia, or Women and Knowledge* (London: Kegan Paul, Trench, Trubner, 1925) casts a liberal feminist position on motherhood in the discourse of scientific meliorism.

14. Jeffrey Weeks, *Sex, Politics, and Society: The Regulation of Sexuality since 1800*, 2d ed. (London: Longman, 1989), 128–131.

15. Leonore Tiefer, "A Feminist Perspective on Sexology and Sexuality," in *Feminist Thought and the Structure of Knowledge*, ed. Mary McCanney Gergen (New York: NYU Press, 1988), 16–26, 22.

16. Weeks, *Sex, Politics, and Society*, 185, citing a passage from the proceedings of the Second International Congress for the World League for Sexual Reform, which took place in 1928.

17. "Its specific planks included support for the political, economic and sexual equality of women and men; reform of marriage and divorce laws; improved sex education; the control of conception; reform of the abortion laws; the prevention of venereal disease and prostitution; the protection of unmarried mothers and the illegitimate child; and the development of rational attitudes towards sexual 'abnormality.'" Ibid., 185.

18. Norman Haire, ed., *Sexual Reform Congress* (see intro., n. 41).

19. Sheila Jeffreys has observed that the Congress privileged heterosexuality over homosexuality, and the new, scientistic feminism over the prewar feminism of the sexual purity movements. Sheila Jeffreys, *The Spinster and Her Enemies* (London: Pandora Press, 1985), 187.

20. The debate thus embodies Strathern's point that "cultural facts never just replicate themselves—they are forever recontextualised, 'borrowed' indeed, and perhaps the repetitions will have the effect of reproducing substance in ways that never take quite the same form." Marilyn Strathern, *Reproducing the Future*, acknowledgments (see intro., n. 49).

21. Rita Felski, "Modernism and Modernity: Engendering Literary History," in *Rereading Modernism: New Directions in Feminist Criticism*, ed. Lisa Rado (New York: Garland Press, 1994), 191–208, 192. Andreas Huyssen has done pathbreaking work in identifying debts within postmodernism to the culture of modernity, and John Christie has continued that project in his revisionist reading of the oppressive modernist roots of Donna Haraway's emancipatory image of the cyborg. Huyssen, *After the Great Divide: Modernism, Mass Culture, Postmodernism* (Bloomington: Indiana University Press, 1986), and Christie, "A Tragedy for Cyborgs," *Configurations* 1 (winter 1993): 171–196.

22. Haldane's list of biological innovations ranges from the unarguable [domestication of plants and animals, "bactericide," contraception] to the disturbingly racist (a change in our "idea of beauty from the steatopygous Hottentot to the modern European"). *Daedalus*, 43.

23. Haldane's fantasy of extracting ova from the dead returned, with a bizarre twist, in 1994, when it was reported that the British Medical Association had been asked to approve "a method of producing test-tube babies using eggs from aborted fetuses." Dr. Roger Gosden of the Edinburgh Medical

School, who announced success in harvesting the ova of laboratory mice, claimed that he is "certain the technique can be transferred to humans." As if confirming Strathern's analysis, the response to proposed reproductive innovations emphasized the disruption of existing kinship relations: "'We are creating some nightmare scenarios in the next century,' said Anthony Clare, a psychiatrist. 'People will be growing up whose mothers were aborted fetuses.'" "Creating Babies From the Unborn?" *Newsday*, 4 January 1994, 1, 25.

24. Edward Yoxen, "Historical Perspectives on Human Embryo Research" (see Chap. 1, n. 50). Heape was not only the factual father of the fictional ectogenesis; he was also a noted author of antifeminist tracts. *Sex Antagonism* (1913) is an analysis of the feminist movement that ascribes its growth to sexual frustration rather than social oppression. It is ironic, even ominous, that the "father" of ectogenesis should have proselytized for a biologically determined, apolitical view of woman. Jeffreys, *The Spinster and Her Enemies*.

25. Bjorn Westin, Rune Nyberg, and Goran Enhorning, "A technique for the perfusion of the previable human fetus," *Acta Paediatrica*, no. 47 (1958): 339–349; J. C. Callahan et al., "Study of prepulmonary bypass in the development of an artificial placenta for prematurity and respiratory distress syndrome of the newborn," *Journal of Thoracic and Cardiovascular Surgery* 44, no. 5 (1962): 600–607; J. C. Callahan, Earl A. Maynes, and Henry R. Hug, "Studies on lambs of the development of an artificial placenta," *Canadian Journal of Surgery* 8 (1965): 208–213; Kermit E. Krantz, Theodore C. Panos, and James Evans, "Physiology of maternal-fetal relationship through extracorporeal circulation of the human placenta," *American Journal of Gynecology* 86, no. 9 (1962): 1214–1288; Warren M. Zapol, Theodor Kolobow, Joseph E. Pierce, Gerald G. Vurex, and Robert L. Bowman, "Artificial placenta: Two days of total extrauterine support of the isolated premature lamb fetus," *Science*, 31 October 1969, 617–618; all cited in Corea, *The Mother Machine*, 258–259. See also Kevles, *In the Name of Eugenics*, 189; Michelle Stanworth, *Reproductive Technologies*, 19, 124–125 (see Chap. 1, no. 65); *Parliamentary Debates*, Commons, 6th ser., vol. 174 (1990), cols. 941–942; "Twins born to 59-year-old woman stir ethics controversy," *Globe and Mail* (Toronto), 28 December 1993, A8; William E. Schmidt, "Birth to 59-Year-Old Raises British Ethical Storm," *New York Times*, 29 December 1993, 1, A6; "Modern Maternity," *New York Times*, 3 January 1994, A22; Linda Wolfe, "And Baby Makes 3, Even if You're Gray," *New York Times*, 4 January 1994, A15; Alan Riding, "French Government Proposes Ban on Pregnancies After Menopause," *New York Times*, 5 January 1994, A6.

26. For analyses of the omissions and repressions in such a progressivist, internalist narrative, see Bruno Latour and Steve Woolgar, *Laboratory Life* (see intro., n. 22), and Sandra Harding, *The Science Question in Feminism*, especially chapter 8, "'The Birth of Modern Science' as a Text: Internalist and Externalist Stories," 197–215 (see intro., n. 24).

27. Harding, *The Science Question in Feminism*, 208.

28. As Daniel Kevles defines these terms, "'positive eugenics' . . . aimed to foster more prolific breeding among the socially meritorious, and 'negative

eugenics' . . . intended to encourage the socially disadvantaged to breed less—or better yet, not at all." *In the Name of Eugenics*, 85.

29. For a further discussion of Haldane's socialist commitments, see Gary Werskey, *The Visible College* (intro., n. 36) and Ronald Clark, *J.B.S.: The Life and Work of J.B.S. Haldane* (London: Hodder & Stoughton, 1969). Haldane's posthumously published science fiction novel, *The Man With Two Memories*, continued his exploration of reproductive alternatives, this time in relation to the complexity and multiplicity of human identity. A sample passage, in which the protagonist from far in the future describes his origins, will give the flavor of the novel, which anticipates contemporary varieties of reproductive technology remarkably: "Every human being so far has been born, if you count removal from the womb by operation as birth. But most people in the society from which I came were not born. About a third were born as a result of the sexual intercourse of two parents. A tiny fraction was born of women artificially inseminated by seed of a long dead father which had been preserved at liquid helium temperatures. One or two in a generation were produced parthenogenetically. About two thirds of all children were produced clonally from cells of other persons which were induced to divide and form an embryo. . . . The technique of inducing embryo formation by human cells is rather tricky, and the proper nutrition of the embryo even more so. In fact our species only mastered it about the year 40,000, and minor improvements were still being made in my own time. Indeed we quite expected one or two per cent of embryos produced in this way to die before they could breathe, or develop so abnormally that they had to be destroyed." J.B.S. Haldane, *The Man with Two Memories* (London: Merlin Press, 1976), 19.

30. *Times Literary Supplement*, 19 December 1924, 875.

31. See Huyssen, *After the Great Divide*, for a valuable analysis of this linkage between modernism, mass culture, and the feminine.

32. Haldane takes the use of infant formula as a classic example of the liberating process of cultural naturalization: "Consider so simple and time-honoured a process as the milking of a cow. The milk, which should have been an intimate and almost sacramental bond between mother and child, is elicited by the deft fingers of a milk-maid. . . . No less disgusting a priori is the process of corruption which yields our wine and beer. But in actual fact the processes of milking and of the making and drinking beer appear to us profoundly natural; they have even tended to develop a ritual of their own whose infraction nowadays has a certain air of impropriety." *Daedalus*, 44–46.

In contrast, Ludovici sees it as an oppressive outgrowth of feminism: "The vast multiplication in recent years of patent infant-foods and preparations of cow's milk sufficiently demonstrates the extent to which modern women are failing in this respect. . . . The enormous popularity of artificial feeding . . . must be due to the increased activities of women of all classes outside the home, which is one of the most noticeable features of the Women's Movement." *Lysistrata*, 66–67.

33. Huyssen's original phrase reads: "The problem is not the desire to differentiate between forms of high art and depraved forms of mass culture and

its co-optations. The problem is rather the persistent gendering as feminine of that which is devalued." *After the Great Divide*, 53.

34. Norman Haire, *Hymen, or the Future of Marriage* (London: Kegan Paul, Trench, Trubner, 1927), 5.

35. Eden Paul, *Chronos, or the Future of the Family* (London: Kegan Paul, Trench, Trubner, 1930); Vera Brittain, *Halcyon, or the Future of Monogamy* (London: Kegan Paul, Trench, Trubner, 1929); and J. D. Bernal, *The World, the Flesh, and the Devil* (see intro., n. 42).

36. Weeks, *Sex, Politics, and Society*, 137.

37. Eden Paul, *Socialism and Eugenics* (Manchester: National Labour Press, [1912?]), 14, 16; cited in Weeks, *Sex, Politics, and Society*, 137, 140.

38. Russell, *Hypatia*, 36.

39. For a discussion of the relation between Woolf's use of the kingfisher motif in *Orlando* and its appearance in "The Wasteland," see my essay, "Virginia Woolf's London and the Feminist Revision of Modernism," in *City Images: Perspectives from Literature, Philosophy, and Film*, ed. Mary Ann Caws (New York: Gordon & Breach, 1991), 99–119.

40. Stephen Heath, *The Sexual Fix* (New York: Schocken, 1984), 3. Of course to indict the sexologists for helping in this system of commodity production is not to deny their achievements in other areas. As Heath acknowledges: "To say that liberation is the definition of a new mode of conformity has to be accompanied by the recognition nevertheless of what has been in many respects a truly liberating process: we do, clearly, live in a dramatically improved sexual freedom when compared with our Victorian ancestors; from the introduction of and the provision of access to effective forms of contraception to the general availability of sexual information (in which sexology has indeed been important) there is a range of factors that do make for a genuine liberation." *Sexual Fix*, 3.

41. For essays in which Haraway explicitly negotiates the difficult terrain between wholesale rejection of science as masculinist and oppressive, and wholesale acceptance of science as a tool of power/knowledge, see "Situated Knowledges: The Science Question in Feminism and the Privilege of Partial Perspective," in *Simians, Cyborgs, and Women*, 183–201 (see intro., n. 8), and "Otherworldly Conversations; Terran Topics; Local Terms," *Science as Culture* 3, part 1, no. 14 (1992): 64–98.

42. Maggie Kirkman and Linda Kirkman, *My Sister's Child: Their Own Story* (Ringwood, Victoria: Penguin Books, 1988).

43. Naomi Haldane Mitchison, review of *Halcyon, or the Future of Monogamy*, by Vera Brittain, *Eugenics Review* 21, n.s. (April 1929–January 1930): 300.

44. The comment reverberates forward to Baroness Mary Warnock's choice of the same language in her response to the Human Fertilisation and Embryology Bill: "It would be paradoxical if a democratic and increasingly educated people should reject research and 'put ourselves back into the seventeenth century.' Then, the question of whether or not Galileo and Descartes might pursue and publish their scientific findings was regulated by religious,

not scientific, consideration. We cannot undo the Enlightenment and it would be morally wrong to place obstacles, derived from beliefs not very widely shared, in the path of science." *Times* (London), 7 December 1989.

45. Les Levidow and Kevin Robins, *Cyborg Worlds: The Military Information Society* (London: Free Association Books, 1989), 169; and Stanley Aronowitz, "The Production of Scientific Knowledge: Science, Ideology, and Marxism," in *Marxism and the Interpretation of Culture*, ed. Cary Nelson and Lawrence Grossberg (Urbana: University of Illinois Press, 1988), 531.

46. Jane Gallop, *Around 1981*, 7 (see intro., n. 26). For the use of this term, Gallop acknowledges Gayatri Spivak's essay, "French Feminism in an International Frame," *Yale French Studies* 62 (1981): 177.

47. Gallop, *Around 1981*, 7.

48. Lord Ritchie-Calder, "Portrait of J.B.S. Haldane," *Listener*, 2 November 1967, 565.

49. Ibid., 565–568; Clark, *J.B.S.*, 17–18.

50. J.B.S. Haldane, *Possible Worlds and Other Essays* (London: Chatto & Windus, 1928), 107–119.

51. Krishna R. Dronamraju, *Haldane: The Life and Work of J.B.S. Haldane with Special Reference to India* (Aberdeen: Aberdeen University Press, 1985), 58.

52. Robert Edwards and Patrick Steptoe, *A Matter of Life*, 50 (see intro., n. 27).

53. J.B.S. Haldane, *The Inequality of Man and Other Essays* (London: Chatto & Windus, 1932), 256–257.

54. For the complete text of the poem, see Clark, *J.B.S.*, 257–258. See also Ritchie-Calder, "Portrait," 567.

55. Charlotte Haldane, "My Husband the Professor," 11 September 1965, BBC Third Programme, National Sound Archive M516W, London, England.

56. Clark, *J.B.S.*, 73; J.B.S. Haldane and Julian Huxley, *Animal Biology* (Oxford: Clarendon Press, 1927). "Portrait of J.B.S. Haldane," National Sound Archive, T137R, BBC Third Programme, November 1966. When this programme was published in the *Listener*, the second half of Case's comment was deleted, with the result that Haldane's identification with Otherness seemed less quixotic and more heroic. Ritchie-Calder, "Portrait," 567.

57. Yaeger is here drawing on Mary O'Brien's monumental *Politics of Reproduction*. Patricia Yaeger, "The Poetics of Birth," in *Discourses of Sexuality: From Aristotle to Aids*, ed. Domna Stanton (Ann Arbor: University of Michigan Press, 1992), 285. Mary O'Brien, *The Politics of Reproduction* (Boston: Routledge & Kegan Paul, 1981).

58. Carole Stabile has made a similar point about the effect of the new fetal visualization technology, which erases the pregnant woman and thus "renders female and male contributions to reproduction equivalent." Stabile, "Shooting the Mother: Fetal Photography and the Politics of Disappearance," *camera obscura: A Journal of Feminism and Film Theory* 28 (January 1992): 196.

59. As Ruth Bleier has summed it up: "While the dominant and changing scientific trends in thinking can be painted in broad sweeps coinciding with

particular changing historical eras—the Reformation, the Renaissance, the Scientific Revolution, the Industrial Revolution—one unchanging feature of our Western history . . . is that all the dominant cultures have been patriarchal, whether enlightened, reformed, feudal, capitalist or socialist. Science, like all culture, reflects that consistent historical bias." Bleier, *Science and Gender: Its Critique of Biology and Its Theories on Women* (New York: Pergamon Press, 1984), 2. Ornella Moscucci makes a more specific point about the science of gynecology: "A deeply entrenched belief in our culture holds that sex and reproduction are more fundamental to woman's than to man's nature. . . . This difference is used to prescribe very different roles for women and men. . . . Since the beginning of the nineteenth century, the science of gynaecology has legitimated these views." Moscucci, *The Science of Woman: Gynaecology and Gender in England, 1800–1929* (Cambridge: Cambridge University Press, 1990), 2. See also Londa Schiebinger, *The Mind Has No Sex?* (see Chap. 1, n. 49).

60. Yeager continues, "If the cyborg becomes one site where the boundaries among humans, animals and machines can be 'thoroughly breached' might not the pregnant woman's body offer another site of utopian monstrousness?" "Poetics of Birth," 294.

61. Donna Haraway, "A Manifesto for Cyborgs," 8 (see intro., n. 44). See also "A Cyborg Manifesto: Science, Technology, and Socialist-Feminism in the Late Twentieth Century," in *Simians, Cyborgs, and Women: The Reinvention of Nature*, 150.

62. Haraway, *Simians, Cyborgs, and Women*, 162. As Haraway famously predicts, "A cyborg world might be about lived social and bodily realities in which people are not afraid of their joint kinship with animals and machines, not afraid of permanently partial identities and contradictory standpoints." *Simians, Cyborgs, and Women*, 154.

63. Haraway, *Simians, Cyborgs, and Women*, 150–151, and "Manifesto for Cyborgs," 36.

64. See Craig Owens, "The Discourse of Others: Feminists and Postmodernism," in *Postmodern Culture*, ed. Hal Foster (London: Pluto Press, 1983), 57–82.

65. Christie, "A Tragedy for Cyborgs," 173.

66. That anti-abortion rhetoric has been gradually driving a wedge between the fetus and the pregnant woman is by now a familiar observation in feminist circles. For more on this topic, see Corea, *The Mother Machine*; Rosalind Pollack Petchesky, "Foetal Images: The Power of Visual Culture in the Politics of Reproduction," in *Reproductive Technologies: Gender, Motherhood, and Medicine*, ed. Michelle Stanworth (Minneapolis: University of Minnesota Press, 1987), 57–80; and Yeager, "Poetics of Birth." A Toyota advertisement appearing in *The Australian*, 15 March 1993, 9, figured the fetus-as-passenger. I am grateful to Jane O'Sullivan, of the Department of English, University of Newcastle, New South Wales, Australia, for sending me this clipping.

67. Sabra Chartrand, "Patents," *New York Times*, 19 July 1993, D2.

68. Ibid. In what seems another instance of fiction setting the agenda for

science, the plans for this ectogenetic chamber (patented in 1991) are an uncanny echo of Robin Cook's fictional description of a similar (but larger) chamber in his 1989 novel *Mutation* (which I discuss in the Introduction).

69. Dr. William Cooper, "Placental Chamber—Artificial Uterus," United States Patent number 5,218,958, filed 21 February 1991; granted 15 June 1993; column 1.

70. The very word choice here participates in the discursive coding of anti-abortion rhetoric: the birth boundary is ignored, and the gestating fetus is referred to as "the baby." Dr. Cooper's anti-abortion construction of ectogenesis echoes Peter Singer and Deane Wells: "Ectogenesis conceivably could win the support of right-to-life organizations and others opposed to abortion. . . . Abortions would in effect become early births, and the destruction of the unborn would cease." *Making Babies*, 119.

71. Gina Kolata, "Operating on the Unborn," *New York Times Magazine*, 14 May 1989, 35. Don Terry, "A Child Is Born in Court Case over Caesarean," *New York Times*, 31 December 1993, A12.

3 SEX SELECTION, INTERSEXUALITY,
AND THE DOUBLE BIND OF FEMALE MODERNISM

The epigraph to this chapter is quoted from Les Levidow, "Sex Selection in India," 151 (see note 4).

1. *Times* (London), 22 April 1990, A11.

2. Sharon Kingman, Steve Connor, and Nicholas Comfort, "Inside Story: Human Embryo Research The Bitter Fight about the Meaning of Life," *Independent*, 22 April 1990, 17.

3. *Times* (London), 8 April 1990.

4. They range from amniocentesis and ultrasonography to chorionic villus biopsy and embryo biopsy. For a critical discussion of some of these techniques, see Les Levidow, "Sex Selection in India: Girls as a Bad Investment," *Science as Culture* 1, no. 1 (1987): 141–152; see also Vibhuti Patel, "Sex-Determination and Sex Pre-Selection Tests in India: Recent Techniques in Femicide," *Reproductive and Genetic Engineering* 2, no. 2 (1989): 111–120; Lakshmi Lingam, "New Reproductive Technologies in India: A Print Media Analysis," *Issues in Reproductive and Genetic Engineering* 3, no. 1 (1990): 13–22.

5. Charlotte Haldane, *Truth Will Out* (London: George Weidenfeld & Nicolson, 1949), 15.

6. Charlotte Haldane, *Man's World* (London: Chatto & Windus, 1926). Peter Firchow mentions Haldane's novel as one of the sources for Aldous Huxley's *Brave New World* in "Science and Conscience in Huxley's *Brave New World*" (see Chap. 1, n. 56).

7. Of course, the designations "superior" and "inferior" were not only suffused with ideology but self-confirming, for they produced a society reinforcing their narrow constructions. See Daniel J. Kevles, *In the Name of Eugenics* (see intro., n. 35), and Les Levidow, "IQ as Ideological Reality," in *Radical Science: Essays*, ed. Levidow (London: Free Association Books, 1986), 198–220.

8. Magnus Hirschfeld, "Presidential Address: The Development and

Scope of Sexology," in *Sexual Reform Congress*, ed. Norman Haire, xiv (see intro., n. 41). See also Jeffrey Weeks, *Sexuality and Its Discontents* (London: Routledge, 1985), 64.

9. Hirschfeld, "Presidential Address," xii. Hirschfeld claimed eugenics as another branch of sexology, and observed that "persons whose impulse departs very widely from the normal must be regarded as unsuited either for marriage or for reproduction." "Presidential Address," xiii. This conference marked the "high point" of the field of sexology, which after 1930 suffered from the combined effects of the rise of fascism, the economic depression, and the drift to war in Europe. Sheila Jeffreys, *The Spinster and Her Enemies*, 186–187 (see Chap. 2, n. 19).

10. Weeks, *Sexuality and Its Discontents*, 67–69; Jeffrey Weeks, *Sex, Politics, and Society*, 104–117 (see Chap. 2, n. 14). Anne Fausto-Sterling offers the following terms, in an attempt to categorize the range of intersex positions: *herms*, "the so-called true hermaphrodites . . . who possess one testis and one ovary"; *merms*, or "male pseudo-hermaphrodites . . . who have testes and some aspects of female genitalia but no ovaries"; and *ferms*, "female pseudo-hermaphrodites . . . who have ovaries and some aspects of the male genitalia but lack testes." Although Fausto-Sterling's contention—that intersexuality is not a disease, and should not be treated as such—is a welcome corrective to a pervasive medicalization of human sexuality and reproductivity, there are problems with her approach. She quite rightly comments on the human tendency to police sex and gender boundaries, seeing it as a reaction to our anxiety about liminality and uncategorizability. Yet in her new categories for intersexuality—herms, merms, and ferms—she reinstitutes the very boundary-drawing practice that her article began by critiquing. "How Many Sexes Are There?" *New York Times*, 12 March 1993, A15. My thanks to Deborah Davenport, M.D., for advice on the nature of the intersex and on sex determination in general.

11. For example, in the classic biology textbook *The Science of Life*, Julian Huxley and his coauthors described the intersex as a literalization *on the level of biology* of last stages of the sex war, not as a victory but as an exhausted truce: "Beneath the calm orderly development of the individual there is war in the nuclei, and in all of us, male or female, the rival forces are present, one outnumbered by the other. In rare cases the battle nearly results in a draw, and that draw marks itself in indeterminate structure. And even when there is a definite win the defeated side may regain a little ground." H. G. Wells, Julian Huxley, and G. P. Wells, *The Science of Life* (London: Cassell & Company, 1931), 638. See also Weeks, *Sexuality and Its Discontents*, 87–88.

12. Wells, Huxley, and Wells, *The Science of Life*, 639. As Wells, Huxley, and Wells point out, the meaning of intersexuality is not innate, but culturally determined, shaped by the values of the culture in which the biological anomaly surfaces. An example of this that they cite occurs among the intersexual pigs abundant in the New Hebrides: "If in Britain the acceptance of a knighthood were necessarily bound up with the ritual slaughter of a number of intersexual pigs, there would be a great demand for sows which produced such valu-

able monstrosities. . . . Sows carrying this gene [for intersexuality] . . . are eagerly sought after and command high prices. As a result of this artificial selection, the proportion of intersexes has grown until it is hundreds of times greater than in the pigs of any other country. This is one of the prettiest examples of the power of selection, given an initial variation, to build up fantastic types." *The Science of Life*, 638.

13. Sheila Jeffreys discusses these notions in *The Spinster and Her Enemies*, 129.

14. Weeks, *Sex, Politics, and Society*, 128–131; Jeffreys, *The Spinster and Her Enemies*, 134–136. Angela Ingram examines the dangerous political consequences of the notion of vocational motherhood in "Un/Reproductions: Estates of Banishment in English Fiction after the Great War," in *Women's Writing in Exile*, ed. Mary Lynn Broe and Angela Ingram (Chapel Hill: University of North Carolina Press, 1989), 326–348.

15. Frank Mort, *Dangerous Sexualities: Medico-Moral Politics in England since 1830* (London: Routledge & Kegan Paul, 1987), 97, and Darko Suvin, *Metamorphoses of Science Fiction: On the Poetics and History of a Literary Genre* (New Haven: Yale University Press, 1979), 103.

16. Among the important research being done by early-twentieth-century biologists and embryologists was Jacques Loeb's discovery of the technique for mechanically induced parthenogenesis in sea urchins (1899), Alexis Carrel's development of a technique for maintaining frog embryo nerve tissues alive outside the body (1907), W. E. Castle and John C. Phillips's work transplanting the ovaries of guinea pigs (1909), and Hans Spemann's discovery of techniques for embryonic grafting (1918). Gordon Rattray Taylor, *The Science of Life*, 255–262 (see Chap. 1, n. 53). See also Mordecai L. Gabriel and Seymour Fogel, eds., *Great Experiments in Biology* (Englewood Cliffs, N.J.: Prentice-Hall, 1955), 204–208.

17. Charlotte Burghes, "The Sex of Your Child," *Daily Express* (London), 18 May 1924, 7. Although her name by her first marriage was Burghes, for clarity and convenience I refer to her as Charlotte Haldane both before and after her second marriage to J.B.S. Haldane.

18. In her autobiography, Charlotte Haldane recalls, "Apart from my early acquaintance with the menace of anti-semitism, the second greatest influence on my youthful mind was that of feminism. I cannot remember a time when I did not ardently regret having been born a girl, nor long to have been a boy." *Truth Will Out*, 6.

19. The article predicts, "The question of colonisation will be reviewed in a new light. Colonial nations that cannot at present produce sufficient males to populate their vast territories may then be able to do so." C. Haldane, "The Sex of Your Child," 7.

20. As she argues in *Motherhood*, whereas male homosexuals have played "useful, and often even heroic parts in social and political spheres," the "type of woman [called] the 'intersex' . . . presents modern society with a grave problem." Charlotte Haldane, *Motherhood and Its Enemies* (London: Chatto & Windus, 1927), 146.

21. Jeffreys, *The Spinster and Her Enemies*, 174–175.

22. Jane Lewis, *The Politics of Motherhood: Child and Maternal Welfare in England, 1900–1939* (London: Croom Helm, 1980), 221.

23. This is another explanation for the turn to biological analogy that I examine in my chapter on the uses of analogy by Julian Huxley. Despite its dangers, analogy has its uses as a vehicle with which to negotiate an otherwise impassible discursive terrain. As Lewis continues, "The ideas formulated by scientists and mediated by the medical profession formed the framework within which all women, including active feminists . . . had to work." Jane Lewis, *Women in England, 1870–1950* (Bloomington: Indiana University Press, 1984), 82.

24. Haldane's columns appeared in the *Daily Express* on 2, 10, 14, and 15 July 1924. They appeared in the *Sunday Express* on 13 July 1924. Columns in the *Lancet* appeared on 25 April 1925 and 27 June 1925.

25. The fictional reproductive technologies of Haldane's *Man's World*, Aldous Huxley's *Brave New World*, and Naomi Mitchison's *Solution Three* represent acute extrapolations from ongoing work in animal husbandry. That there are important facilitating interconnections between human and animal reproductive research, connections relying on an analogical relationship between the experimental subjects in both fields, as well as a suppression of their psychological, philosophical, and moral differences, is apparent if we compare the *Daily Express* debate on sex determination with the lead paragraphs of an article appearing in 1989, in the *New Scientist*: "Scientists in Australia have developed a portable kit to sex animal embryos and split them to double their numbers—in a three-hour procedure that can take place on the farm. The technique, which supersedes laboratory methods that will take two weeks, will simplify the breeding of cattle and other livestock.

"Trials of the embryo-sexing technique in the laboratory and in the field have shown it to be almost completely accurate. The method, developed by a team at the Australian National University (ANU) in Canberra, enables farmers and breeders to choose the sex of seven-day-old embryos quickly and easily, before they are transplanted into surrogate mothers." Liz Glasgow, "Kit for sexing embryos sets to work down on the farm," *New Scientist*, 9 November 1989, 19.

26. Although this story has no byline, it was probably written by Charlotte Burghes [Haldane], since she had special status by now as the *Sunday Express* correspondent of choice on scientific matters. *Sunday Express*, 13 July 1924, 1.

27. The method built on the notion of a gendered "sidedness" of the body expressed four years earlier in Arabella Kenealy's classic antifeminist work *Feminism and Sex Extinction* (London: T. Fisher Unwin, 1920). I am grateful to Martha Macintyre for lending me her copy of this rare text.

28. Kenealy takes a slightly different position on the notion of gender-based sidedness, seeing the ovum as containing "both male and female factors," which are differentiated in the new individual created, once he or she reaches maturity: "The Germ-Plasm proper is inherent in the ovum, in which

it exists in potential, or undifferentiated, form, and . . . it becomes differentiated (in both sexes) into a right and a left-reproductive gland of contrary sex-influence, by differentiative power of the dual-sexed sperm-cell." Ibid., 310, 311.

29. Anne Fausto-Sterling, "Life in the XY Corral," *Women's Studies International Forum* 12, no. 3 (1989): 322.

30. As Fausto-Sterling points out, "the gossip network . . . attempted to discredit scientist Ruth Sager's ground-breaking work on cytoplasmic inheritance, carried out under difficult circumstances because she was a woman and working on a stigmatized subject, by referring to it as 'Ruth's defense of the egg.'" Ibid.

31. Barbara Katz Rothman, *Recreating Motherhood: Ideology and Technology in a Patriarchal Society* (New York: W. W. Norton, 1989); Rayna Rapp, "Chromosomes and Communication: The Discourse of Genetic Counseling," *Medical Anthropology Quarterly* 5, no. 2 (1988): 143–157; Emily Martin, *The Woman in the Body* (see intro., n. 34). Martin describes the moment of insight that centered her work: "The realization that statements about uterine contractions being involuntary are not brute, final, unquestionable facts but rather cultural organizations of experience came to me as a sudden and complete change of perspective. All at once I saw these 'facts' as themselves standing in need of explanation, the way I nearly always saw statements of 'fact' about the body made by Chinese villagers in the field: this is a 'hot' illness, your yin and yang are out of balance, and so on." *The Woman in the Body*, 10–11. Barbara Duden uses a similar strategy of phenomenological investigation in her historical study, *The Woman Beneath the Skin: A Doctor's Patients in Eighteenth-Century Germany*, trans. Thomas Dunlap (Cambridge: Harvard University Press, 1991).

32. Marianne DeKoven, "Gendered Doubleness and the 'Origins of Modernist Form,'" *Tulsa Studies in Women's Literature* 8 (spring 1989): 19.

33. Rita Felski, "Modernism and Modernity: Engendering Literary History" (see Chap. 2, n. 21).

34. Bonnie Kime Scott, *The Gender of Modernism* (Bloomington: Indiana University Press, 1990), 2.

35. Peter Faulkner's definition of modernism lists the following traits as "central to the Modernist aesthetic": "the ideal of the impersonality of the writer" (Joyce); "acutely rendered material . . . [organized] in an illuminating way," rather than "traditional English reliance on the author's personal feelings or experiences"; "objective—no slither; direct—no excessive use of adjectives, no metaphors that won't permit examination" (Pound, defining Imagism); "disinterestedness"; and a "more detached tone." These attributes of modernism are discursively coded as both scientific and masculine. Peter Faulkner, "Introduction," in *The English Modernist Reader, 1910–1930*, ed. Faulkner (Iowa City: University of Iowa Press, 1986). To my mind, the most vivid example of the masculinist subtext of modernism is Ezra Pound's description of the modern writer's creativity: "driving any new idea into the great passive vulva of London." Ezra Pound, "Postscripts to *The Natural*

Philosophy of Love by Remy de Gourmont," in *Pavannes and Divagations* (New York: New Directions, 1958), 207; cited in Carolyn Burke, "Getting Spliced: Modernism and Sexual Difference," manuscript courtesy of the author.

36. Andreas Huyssen, *After the Great Divide*, 53 (see Chap. 2, n. 21).

37. Suzanne Clark, *Sentimental Modernism* (Bloomington: Indiana University Press, 1991).

38. DeKoven, summarizing Edward Said's analysis of the linkage between the modernist attack on realism and a revolutionary social program. Marianne DeKoven, *Rich and Strange: Gender, History, Modernism* (Princeton: Princeton University Press, 1991), 7. The reference is to Edward Said, *Beginnings: Intention and Method* (Baltimore: Johns Hopkins University Press, 1975).

39. Continuing this reevaluation of the political meaning of stylistic or formal traits when they are used by women writers in the context of modernity, Felski observes: "Women's use of supposedly 'outdated' forms such as the autobiography or the *Bildungsroman* cannot simply be interpreted as anachronistic or regressive; the important and wide-spread reappropriation or reworking of such textual models indicates that the project of modernity is indeed an unfinished history, that concerns with subjectivity and self-emancipation encoded within such narrative structures possesses a continuing and often urgent relevance for oppressed social groups." Rita Felski, *Beyond Feminist Aesthetics: Feminist Literature and Social Change* (Cambridge: Harvard University Press, 1989), 161, 169.

40. John Christie and Sally Shuttleworth, eds., *Nature Transfigured: Science and Literature, 1700–1900* (Manchester: Manchester University Press, 1989), 4–5.

41. Thomas Kuhn, *The Structure of Scientific Revolutions* (see intro., n. 16); Bruno Latour and Steve Woolgar, *Laboratory Life* (see intro., n. 22); Evelyn Fox Keller, *Secrets of Life, Secrets of Death* (see Chap. 1, n. 3); Joseph Rouse, *Knowledge and Power: Toward a Political Philosophy of Science* (Ithaca: Cornell University Press, 1987), especially chapter 4; Londa Schiebinger, *The Mind Has No Sex?* (see Chap. 1, n. 49); Donna Haraway, *Primate Visions* (see intro., n. 14); The Biology and Gender Study Group, "The Importance of Feminist Critique for Contemporary Cell Biology," in *Feminism and Science*, ed. Nancy Tuana (Bloomington: Indiana University Press, 1989).

42. Christie and Shuttleworth, *Nature Transfigured*, 5.

43. DeKoven, "Gendered Doubleness," 19. I depart from DeKoven's more canonical understanding of modernist form, because I am persuaded by the work of Huyssen and others that modernism is a problematically narrow reaction-formation to a debased, feminized mass culture. Nonetheless, I find DeKoven's essay a crucial mapping of the feminist rereading of modernism. Her image of the monstrous childbirth foundational to our customary construction of modernism was one of the germs of this chapter, and it powerfully condenses modernism's problematic beginnings as a splitting and denial of the productive/reproductive social sphere.

44. In her essay *"Frankenstein*: A Feminist Critique of Science," in *One Culture: Essays in Science and Literature*, ed. George Levine (Madison: University of

Wisconsin Press, 1987), 287–312, Anne K. Mellor argues that *Frankenstein* is the first and most powerful feminist critique of modern science. I situate the image of autonomous male birth within the cultural shift from romanticism to postmodernism in my essay, "Reproducing the Posthuman Body: Ectogenetic Fetus, Surrogate Mother, Pregnant Man," in *Posthuman Bodies*, ed. Ira Livingston and Judith Halberstam (Bloomington: Indiana University Press, forthcoming).

45. As I explored in Chapter 2, the image of the cyborg, originating in the ectogenesis debate in the early twentieth century, links the modern expression of this womb envy to its postmodern expression.

46. The phrase "the forgotten vagina" appears in a passage cited by De-Koven from Luce Irigaray's "Plato's *Hystera*." DeKoven's analysis reveals the connection between the modernist literary text and the "culturally suppressed originary birth canal." *Rich and Strange*, 36–37. DeKoven and I differ on the nature of that connection, however. I view the male retreat into modernism as a defensive, anxiety-laden response to feared female power, whereas DeKoven views canonical male modernism as ambivalent about female power rather than reactive against it. She holds that "the paradigm of *sous-rature* was a function of the modernist writers' irresolvable ambivalence toward the possibility of radical social change promised/threatened by . . . the 'revolutionary horizon' of the twentieth century." And she argues that "this ambivalence was differently inflected for male and female modernists. Male modernists generally feared the loss of hegemony the change they desired might entail, while female modernists feared punishment for desiring that utter change." DeKoven, *Rich and Strange*, 4. Clearly, this is a difference of emphasis and degree rather than of substance: both of us would, I think, agree that attraction to female potency is the underside of repressive male modernism, but we differ in how we see male modernists responding to that (repressed/denied) attraction.

47. It also involves a reworking of the *Frankenstein* plot, just as Charlotte Haldane also reverses it in her memoirs when she tells the story of their eventual intellectual collaboration and romantic involvement culminating in marriage. She extends Mary Shelley's science fiction narrative of the monstrous creature managed by his scientist-creator in her explanation of how she catalyzed J.B.S.'s career as a writer of popular science essays. Having read his *Daedalus, or Science and the Future*, she recalls, she foresaw "the demand that would follow, for popular articles by him on scientific themes. Unlike most of his predecessors and contemporaries engaged in scientific research, J.B.S. had an unusual enthusiasm for the popularising of science, and a firm conviction that this was both desirable and necessary. . . . We decided that I would become his secretary and agent." *Truth Will Out*, 21. In a later interview for the BBC, Charlotte elaborates on this memory in ways that recall the romantic roots of science fiction—ranging from *Frankenstein* to Fritz Lang's *Metropolis*—figuring J.B.S. as the scientifically gifted cyborg whose machinelike functions she correctly manipulates: "I began to feed questions into this human computer, and out would come explanations as to how things and

people worked. I then typed it out and proceeded to sell them to various magazines and later to English and American publishers. In two years, I created a legend and doubled J.B.S. Haldane's income." Charlotte Haldane, "My Husband the Professor" (see Chap. 2, n. 55).

48. Janice Radway, "The Utopian Impulse in Popular Literature: Gothic Romances and 'Feminist' Protest," *American Quarterly* 33 (summer 1981): 156.

49. Publisher's advertising for *Man's World*, frontispiece to Charlotte Haldane, *Motherhood and Its Enemies*, and Gerald Gould, "News From Nowhere A Return to Romance," *Daily News*, 6 December 1926, 4. While Gould goes on to assert Nicolette's desire to "mate for love, and [have] a baby when she wants to," the article immediately to the left of Gould's review asserts the deterministic power of female biology, recounting how "every artifice which Zoo keepers have employed to take the dead baby baboon, born a week ago, away from its mother on Monkey Hill has failed." J.E.S., "Mother Baboon's Grief. Her Dead Baby," *Daily News*, 6 December 1926, 4.

50. M. Abercrombie, C. J. Hickman, and M. L. Johnson, *The Penguin Dictionary of Biology*, 129–130 (see Chap. 1, n. 31).

51. Kevles, *In the Name of Eugenics*, 70–71. Julian Huxley's representation of the germ-line theory in *The Science of Life* offers an instance of such unconscious gendering. Although according to Weismann's theory the germ-plasm would be perpetuated in, and through, the bodies of women as well as men, the line-drawing in Huxley's coauthored textbook illustrating how "the germ-plasm in each generation produces bodies (soma)" unmistakably represents three white men. *The Science of Life*, 441.

52. *Newsweek*, 2 August 1933, 12; cited in Kevles, *In the Name of Eugenics*, 117.

53. There is a burgeoning body of feminist scholarship on this issue. Among the best recent works are Evelyn Fox Keller, *Reflections on Gender and Science* (New Haven: Yale University Press, 1985), and *Body/Politics: Women and the Discourses of Science*, ed. Mary Jacobus, Evelyn Fox Keller, and Sally Shuttleworth (New York: Routledge, 1990); Schiebinger, *The Mind Has No Sex?*; and Ludmilla Jordanova, *Sexual Visions: Images of Gender in Science and Medicine between the Eighteenth and Twentieth Centuries* (Brighton: Harvester Wheatsheaf, 1990). Donna Haraway has been an important dissenter from this perspective, making the case most eloquently in *Simians, Cyborgs, and Women*.

54. Like early-twentieth-century sexual hygienists, Haldane's hypothetical state has reconstructed "the sexual" as "the racial" (or, as we would now understand it, pertaining to the human race) by conflating reproductivity with sexuality. Mort, *Dangerous Sexualities*, 186.

55. Here I extend Carole Pateman's analysis of male sex-right as integral to post-Enlightenment liberal society. In Haldane's novel, collaboration with the scientific control of reproduction has reduced woman to experimental subject, thus robbing her of agency, whether biological or social. Carole Pateman, *The Sexual Contract* (see intro., n. 29).

56. *Berliner Tageblatt* and *Labour Magazine*, reprinted as part of the advertising matter for *Man's World*, in Charlotte Haldane, *Motherhood and Its Enemies*.

57. The scene contrasts strikingly with Aldous Huxley's ungendered treatment of ectogenesis in *Brave New World*. Despite its horror at the feminized modern industrial world, *Brave New World* does not problematize gender as a category of experience or analysis. Huxley portrays ectogenesis as a process affecting women and men in the same way. Both sexes donate gametes, which the factory combines and modifies, and from which it produces babies to standardized and factory-generated specifications. Huxley's vision is gender-uniform and uniformly dismal: women and men alike are debased by the factory method, as biological mass production serves mass culture.

58. In 1937, Katharine Burdekin memorably fictionalized the Nazi reproductive reduction in *Swastika Night*, ed. Daphnae Patai (New York: Feminist Press, 1985). Elizabeth Russell discusses the parallels between Haldane's and Burdekin's dystopian visions of societies in which men have ultimate control over women's [reproductive] bodies. Elizabeth Russell, "The Loss of the Feminine Principle in Charlotte Haldane's *Man's World* and Katherine Burdekin's *Swastika Night*," in *Where No Man Has Gone Before: Women and Science Fiction*, ed. Lucie Armitt (London: Routledge, 1991), 15–18. See also Nan Bowman Albinski, *Women's Utopias in British and American Fiction* (London: Routledge, 1988).

59. I formulate these qualities of modernism in my essay "Virginia Woolf's London and the Feminist Revision of Modernism" (see Chap. 2, n. 39).

60. Critics disagree on whether Haldane's novel is utopian or dystopian. I question the very possibility of reconstructing a unitary authorial intention for *Man's World*, given the double bind inherent in Haldane's position as a female modernist. To my way of thinking, a more useful approach to the novel is to consider it as a record of the conflicts in Haldane's response, as a feminist, to the dominant discourse of modern science.

61. Between 1934 and 1944 Haldane worked as an undercover agent for the Communist International Brigade (in the Spanish Civil War) and as a journalist in the Soviet Union during World War II. See *Truth Will Out*, especially chapter 7, "Communist International."

62. Like Haldane's writings before it, *Woman Today* is (as its very title indicates) a battleground for different constructions of woman. The inaugural issue of the journal under Charlotte Haldane's editorship carries congratulatory messages from a number of women who would have been anathema to the Charlotte Haldane of *Motherhood and Its Enemies*, in particular Miss Florence White, of the National Spinster's Pensions Association, and Rebecca West, whose endorsement lauds Haldane herself "as one of the finest figures that the woman's movement has brought forward" (March 1939). Later, the distinct feminist emphasis in the fiction and many of the features (such as Sylvia Townsend Warner's series on women in history) warred with the complacent construction of women as proud wives of brave men. My discussion of

Woman Today is based on research in the Haldane archives of the D.M.S. Watson Library, University College, London, and on the microfilm copies of *Woman Today*, held at the British Library (Colindale).

63. *Woman Today*, October 1936, 4, 10, 12; March 1937, 6, 14; April 1937, 5, 12; June 1937, 6, 12; September 1937, 5, 8, 9–10, 13–14; January 1939, 1–5, 16–17; February 1939, 3, 9–10, 11, 16–17; July 1939, 2–3, 8–9.

64. *Woman Today*, July 1939, 8–9.

65. Charlotte Haldane, "They Were Two Hours from Death, But I Was Not Afraid: The Inside Story of My Husband's Experiment," *Woman Today*, August 1939, 1–3, 2.

4 EMBRYOS ARE LIKE PHOTOGRAPHIC FILM

The quotation from Sarah Franklin that appears as an epigraph to this chapter can be found in Sarah Franklin, "Postmodern Procreation: Representing Reproductive Practice," *Science as Culture* 3, part 4, no. 17 (1993): 522–561. This issue of *Science as Culture* appeared after *Babies in Bottles* had already gone into production. It offers a number of valuable analyses of the discursive construction of reproductive technologies.

1. Sir Ian Lloyd, *Parliamentary Debates*, Commons, 6th ser., vol. 171 (1990), col. 96.

2. Carole Stabile argues that "visual representations of fetal autonomy" represent a "historically unprecedented" division of woman and fetus: "In terms of visual and reproductive technologies, and the political interests these technologies often serve, what we are witnessing is the result not of a regression, but a progression. . . . This project of disarticulation, which has been underway for at least two decades, can be alternately read as anti-essentialist (insofar as it denies the material specificity of women's bodies) or as a process of humanizing technology, which then figures as the sign of paternalistic intervention." Stabile, "Shooting the Mother," 179–180 (see Chap. 2, n. 58).

3. Rosalind Pollack Petchesky, "Foetal Images" (see Chap. 2, n. 66). See also Stabile, "Shooting the Mother," and E. Ann Kaplan, "Look Who's Talking, Indeed: The Meaning of Fetal Images in Recent USA Visual Culture," in *Motherhood: Ideology, Experience, Agency*, ed. G. Chung, L. Forcey and E. Nakano Glenn (London: Routledge, 1993).

4. Aldous Huxley to Ludwig von Bertalanffy, 3 October 1949, in *The Letters of Aldous Huxley*, ed. Grover Smith (London: Chatto & Windus, 1969), 603.

5. Aldous Huxley, *The Art of Seeing* (London: Chatto & Windus, 1943), vi–vii. See also Sybille Bedford, *Aldous Huxley: A Biography*, vol. 1, *The Apparent Stability* (London: Paladin Books, 1987), 32–34, and Smith, *Letters*, 39.

6. Grover Smith notes: "Huxley hoped to specialize in biology, so as to prepare for the study of medicine. Eton, as he was to recall during a television interview in 1962 . . . offered . . . a good programme of scientific studies, on which he was already launched." Smith, *Letters*, 38 n. 22. Aldous's letter of thanks to Julian is cited in Bedford, *Apparent Stability*, 16; Gervas Huxley's memoirs of Aldous at Hillside appear in *Aldous Huxley: 1894–1963*, ed. Julian Huxley (New York: Harper & Row, 1965), 58. Because Aldous was nearly

blind while living at Cherwell, his residence there had its dangers. In June 1915, as he reported to his father in a letter, he "walked into some nitric acid, which one of Dr. Haldane's assistants had put outside the lab and left in the pathway. It squirted over my foot and leg . . . it was some time later that I discovered, on removing my sock that my heel was stained a brilliant yellow and was beginning to blister." Smith, *Letters*, 71.

7. Naomi Mitchison, in Julian Huxley, ed., *Aldous Huxley: 1894–1963*, 53.

8. In this, Huxley partially anticipates the analysis of Evelyn Fox Keller, who has described the course of science since the seventeenth-century scientific revolution as the project of drawing the veil away from nature's secrets, with nature constituted as a woman. Yet Huxley himself does not articulate the gendered nature of this impulse to make nature visible. As Keller describes it: "In this interpretation, the task of scientific enlightenment—the illumination of the reality behind appearances—is an inversion of surface and interior, an interchange between visible and invisible, that effectively routs the last vestiges of archaic, subterranean female power. . . . Scientific enlightenment is in this sense a drama between visibility and invisibility, light and dark, a drama in need of constant reenactment at ever-receding recesses of nature's secrets." Keller, "Making Gender Visible in the Pursuit of Nature's Secrets," 69–70 (see Chap. 1, n. 72).

9. Aldous Huxley to G. Gidley Robinson, headmaster of Hillside, 15 May 1915, in Smith, *Letters*, 70.

10. Aldous Huxley, "Fifth Philosopher's Song," in *Leda* (London: Chatto & Windus, 1920); reprinted in *Aldous Huxley's Stories, Essays, & Poems* (London: J. M. Dent & Sons, 1937), 410.

11. Paracelsus's formula for creating a human being in the laboratory is cited in Gordon Rattray Taylor, *The Science of Life*, 11 (see Chap. 1, n. 53). In his monumental work, *A History of Embryology*, Aldous and Julian Huxley's contemporary Joseph Needham cited the following recipe for making a homunculus in book 1 of Paracelsus's *Treatise Concerning the Nature of Things*: "Putrefy human semen for 40 days til it moves and is agitated; then feed it with the arcanum of human blood for forty weeks." Needham, *A History of Embryology*, 2d ed., revised with Arthur Hughes (Cambridge: Cambridge University Press, 1959), 83.

12. Needham, *History of Embryology*, 39–40, 43–44. For two contemporary discussions of the Aristotelian construction of fatherhood as the active, and motherhood as the passive, contribution to conception, see Laqueur, *Making Sex*, and Patricia Yaeger, "The Poetics of Birth" (see Chap. 2, n. 57).

13. Joseph Wood Krutch, "An Elegant Futilitarian," *Literary Review*, 4 March 1922, 464; included in Donald Watt, ed., *Aldous Huxley: The Critical Heritage* (London: Routledge & Kegan Paul, 1975), 69.

14. "The precept therefore follows that, for man, the road to the macrocosm lies through the microcosm." Hermann Muller, *Out of the Night*, 23 (see Chap. 1, n. 68).

15. Huxley's image of ectogenesis in *Crome Yellow* (1921) is widely thought to be a possible inspiration for Haldane's *Daedalus, or Science and the Future*.

Peter Edgerley Firchow, *The End of Utopia: A Study of Aldous Huxley's "Brave New World"* (Lewisburg, Pa.: Bucknell University Press, 1984), 40.

16. Aldous Huxley, *Crome Yellow* (London: Chatto & Windus, 1921), 27. Erasmus Darwin, grandfather of both Charles Darwin and Frances Galton, was "one of the most original-minded men in the history of science." Taylor, *The Science of Life*, 133–134. Darwin played a central part in the eighteenth-century debate about the boundaries of literature and science, combining a fascination with the theory of spontaneous generation, a deep interest in vision, and a technological optimism in poetry that "explicitly described nature as a machine, and life as the product of an industrial process." Maureen McNeil, "The Scientific Muse: The Poetry of Erasmus Darwin," in *Languages of Nature: Critical Essays on Science and Literature*, ed. Ludmilla Jordanova (London: Free Association Books, 1986), 201.

17. When Huxley joined the *Realist*'s editorial board in 1928, he became part of an illustrious group of scientists and writers, including his brother Julian Huxley, close family friends J.B.S. Haldane and Naomi (Haldane) Mitchison, writer Rebecca West, anthropologist Bronislaw Malinowski, and demographer A. M. Carr-Saunders, among others. This new publishing venture, generously supported by H. G. Wells and industrial magnate Sir Alfred Mond, later Lord Melchett, chairman of Imperial Chemical Industries (and model for *Brave New World*'s Mustapha Mond), was committed to making scientific information accessible to the lay public and introducing philosophy and the arts to the scientist. "We stand for making the specialist understood," the editor explained in the first issue, "for introducing the laboratorist, who has lived too long with symbols, to letters, and so giving him that important public which has no time to listen to a man who cannot express himself." To that wider public, the *Realist* brought a broad range of scientific and philosophical/aesthetic issues. In addition to the new venture in film theory, its first volume included essays on rejuvenation therapies (by Dr. Norman Haire, a contributor to the ectogenesis debate), the scientific and social meaning of individuality (by Julian Huxley), developments in aviation (by Professor Hyman Levy), architecture, industrial psychology, linguistic anthropology, trade unionism, and religious revivalism (by Vera Brittain's close friend Winifred Holtby), as well as a study of the novel (by Arnold Bennett), a biographical essay on Richard and Cosima Wagner, and Aldous Huxley's own three-part essay "Pascal." Robert Nichols, "The Movies as Medium," *The Realist: A Journal of Scientific Humanism* 1 (April 1929): 144–164.

18. Nichols, "Movies as Medium," 148–149.

19. In this third attribute of the cinema, Nichols anticipated postmodern approaches to the cinema image as historical artifact. As Tim Boon summarizes one such work: "When we replace the naive view that 'the camera never lies,' it is helpful instead to concentrate on the process and circumstances of the images's constitution. These arguments can be applied with equal validity to the moving images of the cinema film as to still photography." Tim Boon, "The Smoke Menace: Cinema, Sponsorship, and the Social Relations of Sci-

ence in 1937," in *Science and Nature: Essays in the History of the Environmental Sciences*, ed. Michael Shortland (Oxford: BSHS, 1993), 60. Boon is summarizing Ludmilla Jordanova, "Medicine and visual culture (review article)," *Social History of Medicine* 3 (1990): 89–99.

20. Linda Williams, "Film Body: An Implantation of Perversions," in *Narrative, Apparatus, Ideology: A Film Theory Reader*, ed. Philip Rosen (New York: Columbia University Press, 1986), 509. See also Giuliana Bruno, "Spectatorial Embodiments: Anatomies of the Visible and the Female Bodyscape," *camera obscura* 28 (January 1992): 238–261.

21. Bernard Doray, *From Taylorism to Fordism*, 79 (see intro., n. 34). Fatimah Tobing Rony, "Those Who Squat and Those Who Sit: The Iconography of Race in the 1895 Films of Felix-Louis Regnault," *camera obscura* 28 (January 1992): 264.

22. Thomas W. Bohn and Richard L. Stromgren, with Daniel H. Johnson, *Light and Shadows: A History of Motion Pictures* (Port Washington, N.Y.: Alfred Publishing, 1975), 12–13.

23. Boon, "Smoke Menace," 66, 64 n. 18. Boon particularly cites Paul Swann, *The British Documentary Film Movement, 1926–1946* (Cambridge: Cambridge University Press, 1989).

24. Boon, "Smoke Menace," 74–75.

25. Tim Boon, "'Lighting the Understanding and Kindling the Heart'"?: Social Hygiene and Propaganda Film in the 1930s" (paper delivered to the Society for the Social History of Medicine, Oxford, 7 September 1989), 2, manuscript courtesy of the author.

26. Tim Boon, abstract for "'Lighting the Understanding and Kindling the Heart'?: Social Hygiene and Propaganda Film in the 1930s," *Social History of Medicine* 3, no. 1 (1990): 140.

27. Boon, "Lighting the Understanding," 5.

28. Although documentary film making was originally associated with the government, being located at the Empire Marketing Board and later at the General Post Office, Tim Boon discerns a shift in documentary style in the mid-1930s that reflects a changed relation to the government and a different construction of the scientific project. "Anxieties about the possibility of scientific amelioration of the human condition, bred by the global depression and widespread unemployment, had led scientists to pull back from the claim of responsibility for the applications of science" into a position of shared citizenship. Boon, "Lighting the Understanding," 6. This movement from expert scientist to concerned citizen parallels the change in how Charlotte Haldane constructed the scientist from *Man's World* to *Woman Today*, which I discussed in Chapter 3.

29. Julian Huxley, *Memories*, 211 (see Chap. 1, n. 12).

30. *From Generation to Generation* (Eugenics Society, 1937). I thank Lesley A. Hall, curator, The Wellcome Institute for the History of Medicine (London), for having this film screened for me. Amy Heckerling, *Look Who's Talking* (Tri-Star Pictures, 1989).

31. Although at least six versions of the Eugenics Society film were made, correspondence between my notes on the film screened at the Wellcome Institute and Boon's summary of "the film as it was seen in 1937" suggests that they were the same. Boon, "Lighting the Understanding," 4.

32. These sequences were shot at the Stoke Park Colony near Bristol, England, a private mental institution whose director R.J.A. Berry, was a member of the Eugenic Society. Ibid., 7.

33. Ibid., 4.

34. C. P. Blacker to Julian S. Huxley, 23 February 1937, Eugenics Society Papers C 186, Wellcome Institute for the History of Medicine.

35. Boon, "Lighting the Understanding," 8.

36. Lesley A. Hall, Wellcome Institute for the History of Medicine, personal communication to the author. The letter, written by Arthur E. Mason on 3 January 1937, asked Huxley, "Would you consent to being the father of my wife's child, possibly by artificial insemmination. [sic] I am writing this as a feeler and purely from the eugenic point of view." Huxley wrote to Blacker in response, "Enclosed a somewhat startling letter—I have written back to say that I am getting in touch with the Eugenics Soc.[sic], as it is important that any such propositions shd [sic] be carefully gone into by a responsible body. What is happening about the eutelegenesis panel of donors. I confess to feeling, in spite of my intellect, an embarrassment in the matter!" Blacker's response, if a bit deficient in the sense of humor Huxley showed, still remarkably anticipated contemporary AID practice: "If I may advise you, I should have no dealings with him along the lines he suggests. If a eutelegenetic service is ever established it seems to be of the first importance that the identity of the eutelegenetic father should not be divulged. One can see hideous complications arising for that individual, especially if the legal father were psychologically abnormal." Eugenics Society Folder, EUG./C. 186, Wellcome Institute for the History of Medicine. See also Kevles, *In the Name of Eugenics*, 353 n. 49 (see intro., n. 35), and Boon, "Lighting the Understanding," for an extended discussion of the legal issues involved in the Society's use of the Mason family in *From Generation to Generation*.

37. Aldous Huxley, *Brave New World* (New York: Harper & Row, 1969), 6.

38. I am thinking here particularly of the lush red imagery of David Cronenberg's *Dead Ringers* (1988) and the street and roof scenes in Ridley Scott's *Blade Runner* (1982).

39. Joseph Needham, "Biology and Mr. Huxley," *Scrutiny*, May 1932, 76–79; reprinted in Watt, *Aldous Huxley*, 204. Rebecca West, writing in the *Daily Telegraph*, anticipated Needham's biologically informed endorsement, and went it one better, faulting Huxley for not drawing more explicit attention to the scientific foundation to the fiction: "If one has a complaint to make against him [Huxley] it is that he does not explain to the reader in a preface or footnotes how much solid justification he has for his horrid visions. It would add to the reader's interest if he knew that when Mr. Huxley depicts the human race as abandoning its viviparous habits and propagating by means of germ

cells surgically removed from the body and fertilised in laboratories (so that the embryo develops in a bottle and is decanted instead of born) he is writing of a possibility that biologists are seeing not more remotely than, let us say, Leonardo da Vinci saw the aeroplane." Rebecca West, "Aldous Huxley on Man's Appalling Future," *Daily Telegraph*, 5 February 1932, 7; reprinted in Watt, *Aldous Huxley*, 198.

40. William Chapman Sharpe, *Unreal Cities: Urban Figuration in Wordsworth, Baudelaire, Whitman, Eliot, and Williams* (Baltimore: The Johns Hopkins University Press, 1990), 63.

41. "To pry an object from its shell, to destroy its aura, is the mark of a perception whose 'sense of the universal equality of things' has increased to such a degree that it extracts it even from a unique object by means of reproduction. Thus is manifested in the field of perception what in the theoretical sphere is noticeable in the increasing importance of statistics. The adjustment of reality to the masses and of the masses to reality is a process of unlimited scope, as much for thinking as for perception." Walter Benjamin, "The Work of Art in the Age of Mechanical Reproduction," in *Illuminations*, ed. Hannah Arendt, trans. Harry Zohn (New York: Schocken Books, 1969), 223.

42. "The magician heals a sick person by the laying on of hands; the surgeon cuts into the patient's body." Ibid., 233.

43. Citing an unspecified work by Durtain, Benjamin observes that surgeons resemble cameramen in that they must use complicated visualization technologies to perform delicate manipulations: "'I refer to the acrobatic tricks of larynx surgery which have to be performed following the reversed picture in the laryngoscope. I might also speak of ear surgery which suggests the precision work of watchmakers!'" Ibid., 248 n. 14.

44. As Benjamin puts it, "For contemporary man the representation of reality by the film is incomparably more significant than that of the painter, since it offers, precisely because of the thoroughgoing permeating of reality with mechanical equipment, an aspect of reality which is free of all equipment. And that is what one is entitled to ask from a work of art." Ibid., 234.

45. Huxley would link innovations in mechanical reproduction to the course of biological reproduction again, in *Beyond the Mexique Bay*. There, he argued that the mass production of literature and art would be accompanied by an inevitable aesthetic decline that would be *population based*: "The population of Western Europe has a little more than doubled during the last century. But the amount of reading—and seeing—matter has increased, I should imagine, at least twenty and possibly fifty or even a hundred times. If there were n men of talent in a population of x millions, there will presumably be 2n men of talent among 2x millions. . . . It still remains true to say that the consumption of reading—and seeing—matter has far outstripped the natural production of gifted writers and draughtsmen." In a curious coincidence, Walter Benjamin turned to this passage from Aldous Huxley's writings to describe how techniques of reproduction lose their "privileged character" once they spread from the elite few to the common many. Benjamin's comment

on this passage is a masterfully understated allusion to the repressive uses of such an interweaving of eugenic and aesthetic assessments of the human collectivity: "This mode of observation is obviously not progressive." Aldous Huxley, *Beyond the Mexique Bay. A Traveller's Journal* (London: Chatto & Windus, 1949), 274 ff.; cited in Benjamin, "Work of Art," 247–248.

46. Charlotte Haldane, writing in *Nature*, dismissed the last chapter of *Brave New World* as "distressing," but she missed the implications of the material that so disturbed her. Playfully suggesting that "Dr. Jekyll and Mr. Hyde are nothing to Dr. Huxley and Mr. Arnold," she argued that Huxley's novel features a pitched battle between the influences of Huxley's grandfather, T. H. Huxley, and his great-grandfather, "Arnold of Rugby." At the conclusion, she maintained, "Dr. Huxley, who knows and cares about biology and music, science and art, is once again ousted by this double of his, morbid, masochistic, medieval-Christian. Mr. Arnold takes charge of the last chapter of *Brave New World . . .*" Yet to define the opposition here as modern science versus medieval Christianity is to miss the ethnic and cultural dimension of the conclusion, which attains its dramatic force by John's positioning (by others and ultimately by himself) as Other. Charlotte Haldane, "Dr. Huxley and Mr. Arnold," *Nature*, 23 April 1932, 597–598; reprinted in Watt, *Aldous Huxley*, 207–209, 209.

47. David King Dunaway, *Huxley in Hollywood* (New York: Harper & Row, 1989), 13, 117–118.

48. As Rony observes, "Cinema . . . is a much less expensive way of circulating nonwestern bodies in 'situ' than is circulating reconstructed 'villages.'" "Iconography of Race," 272.

49. Aldous Huxley to Anita Loos, 13 October, 1945, in Smith, *Letters*, 534–536.

50. This assertion recalls the thesis of Haldane's *Daedalus*: "that the centre of scientific interest lies in biology." Smith, *Letters*, 534; J.B.S. Haldane, *Daedalus*, 10 (see Chap. 2, n. 9).

51. Smith, *Letters*, 534

52. As Ronald Clark reports, Huxley pilloried "J.B.S. as the Shearwater of *Antic Hay*, the biologist forever wrapped in his physiological experiments while friends took his wife to bed." Clark, *J.B.S.*, 57 (see Chap. 2, n. 29).

53. Smith, *Letters*, 535. Compare the conclusion of *Antic Hay*, in which Lancing shows Myra Viveash Shearwater's laboratory, with its "pregnant she-rabbits," its "little black axolotls," "the beetles, who had had their heads cut off and replaced by the heads of other beetles," and a "fifteen-year-old monkey, rejuvenated by the Steinach process." Shearwater, meanwhile, is bicycling furiously in an experimental chamber modeled on those used in the physiological experiments of J.B.S. Haldane. Aldous Huxley, *Antic Hay* (London: Chatto & Windus, 1923), 325–326.

54. In *Ape and Essence* (see Chap. 1, n. 27), Huxley emphasizes how the science that we think of as liberating can also be enslaving, if put to the service of oppressive or barbaric governments. The Camera shows us an army whose prized weapons are biochemical: "On the pressure-tanks of one army are

painted the words SUPER TULAREMIA; on those of their opponents, IM-PROVED GLANDERS, GUARANTEED 99.44% PURE. Each group of technicians is accompanied by its mascot, Louis Pasteur, on a chain." And physics, too, can be of misused for military purposes, as we see in the image of identical, enslaved, Einsteins: first, "there squats Dr Albert Einstein, on a leash, behind a group of baboons in uniform, " then "the Camera moves across a narrow no-man's-land of rubble, broken trees and corpses, and comes to rest on a second group of animals, wearing different decorations and under another flag, but with the same Dr Albert Einstein, on an exactly similar string, squatting at the heels of their jack-boots. Under the tousled aureole of hair, the good, innocent face wears an expression of pained bewilderment. The Camera travels back and forth from Einstein to Einstein. Close shots of the two identical faces, staring wistfully at one another between the polished leather boots of their respective masters." Aldous Huxley, *Ape and Essence*, 33, 32 (see Chap. 1, n. 27). Einstein himself addressed the danger of the misuses of science in an essay, "Why Socialism?" published the same year as *Ape and Essence*. There, he advised, "we should be on our guard not to overestimate science and scientific methods when it is a question of human problems; and we should not assume that experts are the only ones who have a right to express themselves on questions affecting the organisation of society." Albert Einstein, "Why Socialism?" *Monthly Review* (1949); reprinted in *Radical Science: Essays*, ed. Les Levidow (London: Free Association Books, 1986), 215.

55. A television miniseries based on the novel was made in 1978, and finally aired in 1980. Dunaway, *Huxley in Hollywood*, 417 n. 32.

56. As Grover Smith explains in his note to this passage, the Hays Office was "the board of motion-picture censors supported by the American film industry; later the Johnson office." Smith, *Letters*, 535.

57. Huxley, *Art of Seeing*, vii. Huxley's experiences in 1939 with the visual education exercises of Dr. W. H. Bates, as taught to him by Mrs. Margaret D. Corbett formed the basis of the theory in this book, although as he acknowledged in the preface, "I have tried . . . to correlate the methods of visual education with the findings of modern psychology and critical philosophy" (vii).

58. Jerome Meckier argues that in contrast to *Brave New World* and *Island*, which narrate "the attempts of two different societies to progress from the world of time, memory, and anticipation," *Ape and Essence* is "a story of retrogression and recovery." Jerome Meckier, *Aldous Huxley: Satire and Structure* (London: Chatto & Windus, 1971), 190.

59. Peter Firchow, *Aldous Huxley: Satirist and Novelist* (Minneapolis: University of Minnesota Press, 1972), 136.

60. The screenplay begins in a picture palace where "well-dressed baboons" watch the final battle of the human species (29). This scene, in which identical Einsteins battle until all scientists are vanquished by the biological warfare and atomic bomb that are their co-creation, superimposes cinematic production on divine creation-and-destruction, the Producer on God Himself. Huxley, *Ape and Essence*, 28–29.

61. Bohn and Stromgren, *Light and Shadows*, 394–395.

62. Huxley may have gotten this idea from an article by J.B.S. Haldane, according to Dunaway: "If a tenth of a society's members were affected by gamma radiation . . . the species would be doomed . . . because of some unfortunate little accident . . . men of the 50th century may face hare-lipped imbeciles." "Double Crisis," *World Review*, December 1948, 36; cited in Dunaway, *Huxley in Hollywood*, 418 n. 73. Aldous Huxley to Anita Loos, 26 March 1947, in Smith, *Letters*, 569. But it is also a striking reversal of Albert Einstein's observation: "Man acquires at birth, through heredity, a biological constitution which we must consider fixed and unalterable, including the natural urges which are characteristic of the human species. . . . If we ask ourselves how the structure of society and the cultural attitude of man should be changed in order to make human life as satisfying as possible, we should constantly be conscious of the fact that there are certain conditions which we are unable to modify. . . . The biological nature of man is, for all practical purposes, not subject to change." Einstein, "Why Socialism?" 216.

63. Zoe Sofia, "Exterminating Fetuses: Abortion, Disarmament, and the Sexo-Semiotics of Extraterrestrialism," *diacritics* 4 (summer 1984): 47–48.

64. Aldous Huxley, "Stars and the Man," Hubble Collection, Huntington Library; reprinted in Tom Bezzi, *Hubble Time* (San Francisco: Mercury House, 1987), 57.

65. Bezzi, *Hubble Time*, 2.

66. David Abbot, *Astronomers: The Biographical Dictionary of Scientists* (New York: Peter Bedrick Books, 1984), 78; see also Bezzi, *Hubble Time*, 1–3.

67. Abbot, *Astronomers*, 77. Although such a figure might seem out of place in a novel about the film industry, in fact Edwin Hubble was a central member of the Huxleys' close group of friends, most of whom were either expatriate British citizens or film stars, among them Charlie Chaplin and Paulette Goddard, Gerald Heard and Peggy Kiskadden, Charlie MacArthur, Helen Hayes, and Anita Loos, the tiny screenwriter and author of *Gentlemen Prefer Blondes*. See Dunaway, *Huxley in Hollywood*; Bezzi, *Hubble Time*, 1–3; and Bedford, *Apparent Stability*.

68. Aldous Huxley, "Stars and the Man," 60.

69. Robert Edwards and Patrick Steptoe, *A Matter of Life* (see intro., n. 27).

70. See Introduction, "Alternative Visions for Reproduction," pp. 11–13, and "Disjunction in the Feminist Response to RT," pp. 19–23.

71. See Ludmilla Jordanova, *Sexual Visions* (see Chap. 3, n. 53), Londa Schiebinger, *The Mind Has No Sex?* (see Chap. 3, n. 41), and Laqueur, *Making Sex*.

72. Jordanova, *Sexual Visions*, 98, 110.

73. This metaphor of male-male pregnancy has by now become almost a cliché in scientific memoirs. Most recently, it surfaced in what seems initially a surprising place: the memoirs of Carl Djerassi, the synthesizer of the oral contraceptive, or (as he is often labeled) "the father of the Pill." Disputing that label as "phallocentric," Djerassi says that he often points out that "for a new

drug to be born there is needed also a mother and frequently also a midwife or an obstetrician. The organic chemist must first produce the substance; the biologist must then demonstrate its activity in animals; only then can the clinician administer the material to humans. I led the small chemical team at Syntex in Mexico City, which accomplished the first synthesis of a steroid oral contraceptive on 15 October 1951. Gregory Pincus . . . headed the biological group that first reported the ovulation-inhibiting properties of these steroids in animals. The Harvard gynecologist John Rock and his colleagues performed the clinical studies to demonstrate the contraceptive efficacy in humans. If I am the father, Pincus must be the mother, or is it vice versa? At least there is no doubt about the part John Rock played in this birth.

"But to be accurate, one needs to retrace its genealogy at least down to the grandparents and a few uncles." Carl Djerassi, *The Pill, Pigmy Chimps, and Degas' Horse: The Autobiography of Carl Djerassi* (New York: Basic Books, 1992), 49–50.

74. Edwards and Steptoe here recapitulate the process of turning an artifact into a fact, brilliantly anatomized by Bruno Latour and Steve Woolgar, in *Laboratory Life* (see intro., n. 22).

75. A very recent instance of this same RT/VT linkage occurred on 6 July 1993, when the *New York Times* ran a story titled "Miniature Scope Gives the Earliest Pictures of a Developing Embryo." *Times* science reporter Gina Kolata reported the development of an experimental technique, "embryoscopy," that enables a physician to examine a six-week-old embryo "as closely as if he had opened the woman's womb." The narrative of the development of this new technique parallels Patrick Steptoe's development of laparoscopy from the perception that there was a need for the method to the notification of the press once a number of successful procedures had been accomplished. How can an awareness of the historical interconnection of RT and VT help us respond, as feminists, to the representation of this new reproductive technology, as well as to the issues around its actual implementation?

Based on Aldous Huxley's fictions and their influence on the work of Edwards and Steptoe (as well as their earlier influence on Rock and Hertig), we would expect to find that the advance in VT became the site of scientific rivalry, catalyzed advances in RT, and served the unconscious desire to separate the fetus, visually, from the gestating woman. All of these phenomena do indeed seem to characterize the response to the new VT of abdominal embryoscopy: "The next step will be to use embryoscopy to deliver treatment, medical experts said. They could . . . cut amniotic bands, which are constrictions of the amniotic membrane that deform fetuses. Or . . . they may be able to deliver gene therapy. Fetal surgery and fetal therapy in the first trimester is now definitely on the horizon, Dr. Pergament [director of reproductive genetics at Northwestern University in Chicago] said. For now, he added, '*it's a bit of who's going to be the first one to do it*'" (my emphasis). Kolata reports that this new visualization technique can be used diagnostically in cases where there is a high risk of genetic defects in the embryo or fetus. Yet awareness of the way that

VT feeds the need for more VT, and sensitivity to the general principle of "the camera lie," might enable the reader to pick up not only such benign applications for this new technique, but also ones that are more troubling. "Dr. Quintero said he had also been able to reassure some women that their embryos or fetuses were perfectly normal, even though there might have been suspicious shadows on ultrasound." Thus ultrasound technology produces the need for embryoscopy; one can only wonder what technology will meet the need produced by embryoscopy.

76. Catherine Vasseleu, "Life Itself," in *Cartographies: Poststructuralism and the Mapping of Bodies and Spaces*, ed. Rosalyn Diprose and Robyn Ferrell (Melbourne: Allen & Unwin, 1991), 55–64.

77. Kolata, "Miniature Scope," C3.

5 FROM GUINEA PIGS TO CLONE MUMS

1. Sandra Harding, *The Science Question in Feminism*, 21–24 (see intro., n. 24).

2. Donna Haraway has chosen "allegory" to describe a somewhat similar process by which scientific knowledge constructs, as well as reflects, issues that are both physiological and social. "All Western cultural narratives about objectivity are allegories of the ideologies of the relations of what we call mind and body, of distance and responsibility, embedded in the science question in feminism." Donna J. Haraway, *Simians, Cyborgs, and Women*, 190 (see intro., n. 8).

3. As Harding observes: "Science functions primarily as a 'black box': whatever the moral and political values and interests responsible for selecting problems, theories, methods, and interpretations of research, they reappear at the other end of inquiry as the moral and political universe that science projects as natural and thereby helps to legitimate." *The Science Question in Feminism*, 250–251. Evelyn Fox Keller has argued the point extensively in her works, as well. As she expresses it succinctly in a recent essay: "The exclusion of the feminine from science has been historically constitutive of a particular definition of science—as incontrovertibly objective, universal, impersonal—and masculine: a definition that serves simultaneously to demarcate masculine from feminine and scientists from non-scientists—even good science from bad." Evelyn Fox Keller, "The Gender/Science System: Or, Is Sex to Gender as Nature Is to Science?" in *Feminism and Science*, ed. Nancy Tuana (Bloomington: Indiana University Press, 1989), 42.

4. Andrea Henderson, "Doll-Machines and Butcher-Shop Meat," (see intro., n. 31). See also my essay, "Representing the Reproductive Body," *Meridian* 12 (May 1993): 29–45.

5. Carol Gilligan explores the moral implications of reproduction in her book *In a Different Voice: Psychological Theory and Women's Development* (Cambridge: Harvard University Press, 1982). Gilligan's work offers a reevaluation of the Heinz dilemma, central to Lawrence Kohlberg's theory of the six stages of moral development. As Gilligan describes it, "In this particular dilemma, a man named Heinz considers whether or not to steal a drug which he cannot afford to buy in order to save the life of his wife" (25). As Evelyn Fox Keller's

biography of Barbara McClintock suggests, Gilligan's opposition between a female moral structure (emphasizing responsibility and care) and a male moral structure (based on abstract principles of truth and justice) might be transposed to the realm of modern science, where it could suggest the beginnings of a feminist critique of science. Keller emphasizes the constraining power of the discursive community of science, however, which is an improvement on Gilligan's tendency to see moral thinking in essentialist rather than discursively constructed terms. Evelyn Fox Keller, *A Feeling for the Organism: The Life and Work of Barbara McClintock* (New York: W. H. Freeman, 1983).

6. Evelyn Fox Keller, *Secrets of Life, Secrets of Death*, 75 (see Chap. 1, n. 3).

7. Ronald Clark, *J.B.S.*, 31 (see intro., n. 28).

8. Naomi Mitchison, *All Change Here*, 61 (see intro., n. 28).

9. I use the term "gender" to indicate the socially constructed difference between men and women, and their roles, as opposed to the biological difference between men and women; however, that distinction rests upon the assumption that "sex" is natural and only gender is socially constructed, an assumption that does not hold true when we examine the history of sex and sexuality, as Thomas Laqueur among others has demonstrated. I take the position that "sex" (in terms of the biological substrate as well as the practice) is equally socially constructed.

10. I am arguing here that Clark's biography reflects gender-inflected standards of aesthetic coherence similar to those Miller sees as foundational to the novel, standards which construct as aesthetically pleasing, or plausible, plot developments that are in synchrony with our sense of what is gender-appropriate. As she puts it, "The fictions of desire behind the desiderata of fiction are masculine and not universal constructs. It [the blind spot] does not see that the maxims that pass for the truth of human experience and the encoding of that experience in literature are organizations, when they are not fantasies, of the dominant culture." Nancy K. Miller, "Emphasis Added: Plots and Plausibilities in Women's Fiction," *PMLA* 96 (January 1981): 46. Harding, *The Science Question in Feminism*, 63.

11. "On the basis of playing is built the whole of man's experiential existence." D. W. Winnicott, *Playing and Reality* (London: Tavistock Publications, 1971), 75, 48. I disagree with Winnicott, who sees play as a natural phenomenon. I see play as both constructed by culture and constructing identity and culture.

12. Mitchison, *All Change Here*, 61.

13. Keller, *Secrets of Life, Secrets of Death*, 96–97.

14. Keller, *A Feeling for the Organism*, 181.

15. "I squeezed milk from the teats of one of my loved ones, so as to taste it, but it was rather nasty." Mitchison, *All Change Here*, 61. The debates about Gilligan's book have been heated, and there are certainly problems with her model; however, I find her category of moral development a useful conceptual structure with which to examine the meaning of women's tendency to concern ourselves with individual relations more than abstract principles.

16. Naomi Mitchison, "Small Talk . . . Memories of an Edwardian Childhood," in *As It Was: An Autobiography 1897–1918* (Glasgow: Richard Drew Publishing, 1988), 23–24.

17. In *All Change Here* she recalls that because her brother Jack "did not go in for the tobacconist's assistant type who was more or less available . . . up to 1914 his sexual experience, like mine, was practically non-existent, except for erotic verse and guinea pig watching" (74).

18. Naomi Mitchison, *You May Well Ask*, 69 (see intro., n. 37).

19. Ibid., 69–70.

20. Thus after watching guinea pigs with their babies, she concludes, "Females accepted and suckled the babies of those females whom they liked, rejecting those whose mothers they disliked. But is 'like' and 'dislike' the correct way of considering this? There are probably other reasons." Mitchison, *All Change Here*, 62.

21. Daniel J. Kevles, *In the Name of Eugenics*, 58 (see intro., n. 35).

22. Jill Benton, *Naomi Mitchison: A Biography* (London: Pandora, 1990), 23. Benton emphasizes the brother-sister and mother-daughter dynamics in *Saunes Bairos*. I share her interpretation of them, but I find the most significant aspect of the play its critique, from a racially, culturally, and sexually marginal position, of white Western eugenic "science."

23. Sir Ian Lloyd, *Parliamentary Debates*, Commons, 6th ser., vol. 171 (1990), col. 98.

24. Naomi Haldane [Mitchison], *Saunes Bairos: A Study in Recurrence* (Oxford: privately published by the Oxford Preparatory School, 1913). Text courtesy of the author.

25. This reframing echoes that generated by Gilligan's female subjects in their response to the Heinz dilemma: "The proclivity of women to reconstruct hypothetical dilemmas in terms of the real . . . shifts their judgement away from the hierarchical ordering of principles and the formal procedures of decision making." *In a Different Voice*, 100–101.

26. Carole Pateman, *The Sexual Contract*, especially chapter 4, "Genesis, Fathers and the Political Liberty of Sons" (see intro., n. 29).

27. As Daniel Kevles reports, the bill fell short of true eugenic restrictions on reproduction, such as mandatory sterilization. It classified "mental deficiency" according to a functionalist rather than hereditary standard (ranging from "cretinism or mongolism to inability to benefit from education"), and it provided for in-community care, rather than institutionalization, of the mentally handicapped. Kevles, *In the Name of Eugenics*, 98–99.

28. The term "situated knowledges" is Donna Haraway's, although it also appears in the work of Helen Longino. For Haraway: "Even the simplest matters in feminist analysis require contradictory moments and a wariness of their resolution, dialectically or otherwise. 'Situated knowledges' is a shorthand term for this insistence. Situated knowledges build in accountability. . . . Situated knowledges are always *marked* knowledges; they are re-markings, reorientatings, of the great maps that globalized the heterogenous body of the world in the history of masculinist capitalism and colonialism." Donna Hara-

way, "Reading Buchi Emecheta: Contests for 'Women's Experience' in Women's Studies," in *Simians, Cyborgs, and Women*, 111.

Haraway expands on this definition later in the same volume: "Situated knowledges require that the object of knowledge be pictured as an actor and agent, not a screen or a ground or a resource, never finally as slave to the master that closes off the dialectic in his unique agency and authorship of 'objective' knowledge." Donna Haraway, "Situated Knowledges," 199 (see Chap. 2, n. 41).

Helen Longino, commenting on Haraway's work, stresses the communal nature of the feedback processes that negotiate and reconstruct knowledge: "Knowledge is always knowledge in a situation, from a certain point of view. It is, therefore, both incomplete and perspectival. Objectivity is recognition of the local, mediated, situated, and partial character of one's knowledge." Helen Longino, *Science as Social Knowledge: Values and Objectivity in Scientific Inquiry* (Princeton: Princeton University Press, 1990), 212–213.

29. H. G. Wells, Julian Huxley, and G. P. Wells, *The Science of Life*, 222 (see Chap. 3, n. 11).

30. Naomi Mitchison, *Memoirs of a Spacewoman* (see Chap. 1, n. 71).

31. I draw this model of individual versus collective objectivity from Helen Longino, who suggests that we can salvage the notion of scientific objectivity by stressing the social nature of scientific inquiry. Longino, *Science as Social Knowledge*, 62–82. Compare to this Donna Haraway's complex assessment of how "feminists have both selectively and flexibly used and been trapped by two poles of a tempting dichotomy on the question of objectivity." "Situated Knowledges," 183.

32. Here Mitchison anticipates Octavia Butler's play with identity, difference, and cross-species communication in her *Xenogenesis* trilogy, especially its third volume, with its aptly named alien protagonist, Akin. Octavia Butler, *Adulthood Rites* (New York: Warner Books, 1988).

33. Donna J. Haraway, "Otherworldly Conversations; Terran Topics; Local Terms," 92 (see Chap. 2, n. 41).

34. Thomas S. Kuhn, *The Structure of Scientific Revolutions*, 138–139 (see intro., n. 16).

35. Of course, the Aristotelian model of preformation was the notion of the homunculus, which held that the adult was contained with all his parts preformed *in the sperm*. The crucial point, however, is that preformationists held that the embryo already existed, preformed and minuscule, in one of the gametes, whereas the epigenesists held that the embryo developed *de novo* following conception. Gordon Rattray Taylor, *The Science of Life*, 31 (see Chap. 1, n. 53).

36. Ibid., 255.

37. Garland Allen, *Life Science in the Twentieth Century* (New York: John Wiley & Sons, 1975), 31.

38. As Garland Allen observes, "Contemporary wits noted that ever since Loeb's discovery about salt water stimulating parthenogenesis, maiden ladies had expressed grave doubts about ocean bathing; one less subtle observer

noted that weekends at the seashore did indeed seem to be remarkably fertile." Ibid., 77–78.

39. Ibid., 78.

40. Charles Kingsley, *The Water-Babies*, 203 (see Chap. 1, n. 14).

41. Yet Huxley seems to anticipate Mitchison's use of the echinoderm to anchor her revisionist feminist critique of science. In his discussion in *The Science of Life*, he gives us an echinoderm that is neither Promethean nor Epimethean, neither pure prophetic science nor applied technology: it thus arguably represents an entirely new scientific perspective.

42. A later encounter with the echinoderm emphasizes the sheer survival value of such conceptual shifts that decenter the self and take up the perspective of a marginalized Other. In chapter 6 of *Memoirs*, the space scientists are exploring an unfamiliar landscape when Mary "suddenly . . . [realizes] that this landscape of columns and snappers was exactly what one sees under a low power microscope looking at a sea urchin. These hills were simply enormous echinoderms. . . . It was all so obvious. And of course so dangerous" (79). Only when they are able to break free of the earth-centered assumptions about size and scale governing their perceptions, and to identify their surroundings as the surface of a giant echinoderm, are the space explorers able to scramble away from the dangerous grasp of the creature's giant snappers. Their colleagues, unable to make that perceptual shift, are left behind, victim of the category-bound rigid science challenged by Mitchison's *Memoirs of a Spacewoman*.

43. The passage reads as a postmodern, extraterrestrial reframing of Molly Bloom's modernist "yes I said Yes I will yes."

44. Naomi Mitchison, *Comments on Birth Control* (London: Faber & Faber, 1930).

45. In defiance of the binary construction of female sexuality ("baby or not baby"), Mitchison argues that women both want motherhood and its sensuality *and* sexual pleasure: "Part of her tenderness toward her lover expresses itself in the passion to bear him a child, and all contraception is a compromise with this. She gets, even in the moment of supreme sexual enjoyment the sharp flash of longing for the April feel of a baby at her breast." Ibid., 23. The pamphlet was based on "Some Comment on the Use of Contraceptives by Intelligent Persons," Mitchison's contribution to the Third International Congress of the World League for Sexual Reform, and one of only two papers proposing any alternatives to sexual intercourse. The alternative Mitchison proposes appears to be mutual masturbation. Rather than getting into the habit of "more actual copulation than they both really want to the bottom of their hearts," a couple would do better, she argues, to "practice many of the odd forms of sexual enjoyment which come naturally and not at all shockingly to any inventive couple, and which are usually condemned by doctors as 'these practices'" than to risk contraceptive failure early in marriage due to ill-fitting contraceptive methods or a wife's inexperience. Naomi Mitchison, "Some Comment on the Use of Contraceptives by Intelligent Persons," in Norman Haire, ed., *Sexual Reform Congress*, 183 (see intro., n. 41).

46. The reference—and Mitchison makes it explicitly in the text—is to the character of the same name in Shakespeare's comedy of gender trouble, *Twelfth Night*.

47. The implications of such a new vision, as Donna Haraway has observed, are both social and epistemological: "No longer able to sustain the fictions of being either subjects or objects, all the partners in the potent conversations that constitute nature must find a new ground for making meanings together." Haraway, "Otherworldly Conversations," 65.

48. Naomi Mitchison, *Solution Three* (London: Dennis Dobson, 1975), 35, and "The Clone Mums," November 1970, Haldane Papers, National Library of Scotland, Acc. 5831.

49. The draft form of the novel's dedication emphasized the bisexual seductions of science as well as the sexualized nature of her incorporation of Watson's theory: "To Jim Watson who [so to speak] impregnated me with the idea of this book and Anne MacLaren [*sic*] who encouraged it and made it more." Yet in its published version, the dedication emphasizes both the masculine nature of the theory of genetic determinism (suppressing the acknowledgement to MacLaren) and its terrors (describing it as "this horrid idea"). Naomi Mitchison, notebook, "The Clone Mums," November 1970, Haldane Papers. James Watson, *The Double Helix: A Personal Account of the Discovery of the Structure of DNA* (New York: W. W. Norton, 1980). Anne McLaren, a "leading British embryologist" according to the authors of a recent book on reproductive technology, has been a soothing, pro-science voice in the heated debate over reproductive technology. She has gone on record dismissing those who fear the technique of cloning: "As Anne McLaren has pointed out, small numbers of genetically identical people have never been something we have worried about; we don't, after all, think we should kill one of a pair of twins. What frightens critics is not one or two genetically identical individuals, but tens of hundreds, created deliberately to supply some dreadful social need." Linda Birke, Susan Himmelweit, and Gail Vines, *Tomorrow's Child*, 213–214 (see Chap. 1, n. 65).

50. The Biology and Gender Study Group, "The Importance of Feminist Critique for Contemporary Cell Biology," 180 (see Chap. 3, n. 41). See also Mary Jacobus, "Is there a Woman in this Text?" *Reading Women Writing: Essays in Feminist Criticism* (New York: Columbia University Press, 1986), 83–109, and Alan G. Gross, *The Rhetoric of Science*, 54–68 (see intro., n. 23). Evelyn Fox Keller comes up with a curiously opposed formulation: "Ironically enough, DNA is popularly called 'the mother-molecule of life.'" Keller, *Secrets of Life, Secrets of Death*, 51.

51. Watson's assessment of Mitchison in *The Double Helix* suggest that he is impressed not only by her standing as a novelist but also by her family connections. A "distinguished writer," he points out, she was also "a sister of England's most clever and eccentric biologist, J.B.S. Haldane." *The Double Helix*, 63.

52. Biology and Gender Study Group, "Feminist Critique for Contemporary Cell Biology," 180.

53. I am grateful to Waddington's daughter, the mathematician Dusa McDuff, for information about his friendship with Naomi Mitchison, as well as about the egalitarian temper of the Waddington household. See also Biology and Gender Study Group, "Feminist Critique for Contemporary Cell Biology," 180, and Gary Werskey, *The Visible College* (see intro., n. 36).

54. Biology and Gender Study Group, "Feminist Critique for Contemporary Cell Biology," 180. An essay that Waddington wrote in 1969 expanded on this ovum imagery, as part of an argument that "a scientist's metaphysical beliefs are not mere epiphenomena, but have a definite and ascertainable influence on the work he produces." Waddington explains that his decision to become an experimental embryologist originates in part in the "very odd late Hellenistic" notion, taught him by his chemistry master E. J. Holmyard, "which infiltrated into my thinking at a very early stage . . . and which [has] remained there ever since": "*The World egg* 'Things' are essentially eggs-pregnant with God-knows-what. . . . One strand of Gnostic thought asserted that *everything* is like that." C. H. Waddington, "The Practical Consequences of Metaphysical Beliefs on a Biologist's Work: An Autobiographical Note," in *Towards a Theoretical Biology*, vol. 2, *Sketches* (Edinburgh: Edinburgh University Press, 1969), 124. My thanks to Waddington's daughter, Dusa McDuff, for lending me the corrected page proof of this essay.

55. See Kevles, *In the Name of Eugenics*, and Donna Haraway, *Crystals, Fabrics, and Fields*, 133 (see intro., n. 18).

56. In *Solution Three*, Mitchison alternates between capitalizing this word and keeping it in lower case, depending (it seems) on whether she is using it informally to indicate a shared sense of social policy, or formally as a juridical instrument.

57. The reference is clearly to the genetic code contained in the double helix of DNA, which Watson referred to as "the Rosetta stone for unravelling the true secret of life." *The Double Helix*, 18.

58. In its published form, the novel emphasized cloning as the solution to the problems of racism, sexism, and violence. However, while planning the novel Mitchison toyed with the notion of writing about male abdominal pregnancy instead: "Council at work. Discussion. Sex almost always painful sooner or later—Why have it? Woman councillor tells man she dislikes external sex organs. They can discuss the future including the nature of a police state & the crisis of identity. Should there be implants rather than pregnancies? The men might then have them." Mitchison, "The Clone Mums," November 1970, Haldane Papers. Mitchison anticipates the April 1990 discussion paper of Australia's National Bioethics Consultative Committee, which explored the medical and social implications of the hypothetical technique of abdominal pregnancy. William A. W. Walters, "Transsexualism and Abdominal Pregnancy," *Developments in the Health Field with Bioethical Implications* 2 (April 1990).

59. Naomi Mitchison, manuscript of "The Clone Mums," November 1970, Haldane Papers, 1R.

60. Mitchison explored this theme in much greater length in her later

novel, *Not by Bread Alone*, where she linked the genetic engineering schemes of global agribusiness to the struggles of indigenous peoples for self-definition and autonomy. Naomi Mitchison, *Not by Bread Alone* (London: Marion Boyars, 1983).

61. "Was it possible that if they, or any other Clone couple, could see the danger of meiosis, they would consider a fusing technique? In one prepared cell her nucleus could be knocked out; in another, his; then the two could be fused without the uncertainties of meiosis and put back safely. . . . That might indeed produce a child of greater excellence. . . . It need not result in what was, perhaps, the sin of meiosis, if, instead, they would accept fusing. . . . Was this really the solution? . . . For two centuries there had been the anxiety to stabilize: when it was at last achieved that was Solution Three; the world had needed it so desperately. But now there was a counter-danger of subsiding into unquestioning confidence and security in the thought of the Clones taking over. . . . But might not a slightly other kind of excellence be needed, now, having to arrive through—what? A planned accident? In an unplanned plan?" Mitchison, *Solution Three*, 122–123.

62. It also recalls the theory of cytoplasmic influence of Mitchison's good friend, C. H. Waddington, which I discussed earlier in this chapter.

63. Shulamith Firestone, *The Dialectic of Sex*, 206 (see Chap. 1, n. 10). Mapping the goals of a feminist revolution, Firestone's book echoes many of the modernist fantasies for perfecting the human species addressed by participants in the ectogenesis debate (see Chapter 2): "The reproduction of the species by one sex for the benefit of both would be replaced by (at least the option of) artificial reproduction: children would be born to both sexes equally, or independently of either, however one chooses to look at it. . . . The division of labor would be ended by the elimination of labor altogether (cybernation). The tyranny of the biological family would be broken." Firestone, *Dialectic of Sex*, 11. Firestone also anticipates many of the reproductive advances currently debated hotly in the news, among them cloning. Finally, echoing Julian Huxley, she predicted areas of reproductive innovation still to be developed: "Imagine parthenogenesis, virgin birth, as practiced by the greenfly, actually applied to human fertility." *Dialectic of Sex*, 182.

64. Margaret Talbot, "Inside Publishing: A Blast from the Feminist Past," *Lingua Franca*, March/April 1993, 12.

65. For a sample of some of these radical feminist critiques of reproductive technology, see Robyn Rowland, "Of Women Born, but For How Long?" in *Made to Order: The Myth of Reproductive and Genetic Progress*, ed. Patricia Spallone and Deborah Lynn Steinberg (London: Pergamon Press, 1987), 67–83; Gena Corea, *The Mother Machine* (see intro., n. 45); Renate Duelli Klein, *Infertility: Women Speak Out* (London: Pandora Press, 1989); Rita Arditti, Renate Duelli Klein, and Shelley Minden, *Test-Tube Women* (see Chap. 1, n. 65). See also the journal published by Pergamon until 1992: *Reproductive and Genetic Engineering*.

66. Talbot, "Inside Publishing," 13. The 1992–1993 edition of *Books in Print* has no entry for *The Dialectic of Sex*.

67. *Solution Three* will soon be back in print, however. I am currently preparing an edition of it for the Feminist Press (forthcoming, 1995).

68. Haraway, "Otherworldly Conversations," 72. The echo of J.B.S. Haldane's celebrated essay, "Possible Worlds," cannot be accidental. I believe that Haldane is one of the major unacknowledged influences on Haraway's style of popular science writing, particularly in his iconoclastic willingness to embrace different subject positions.

69. Elizabeth Wasserman, "Baby Makers Become Outlaws," *Newsday*, 7 April 1993, 22, 26.

70. Katha Pollit, "When is a Mother not a Mother?" *Nation*, 31 December 1990. See also Christopher Reed, "'Judgement of Solomon' in US Surrogate Mother Case," *The Age* (Melbourne), 24 October 1990, 9.

71. Naomi Mitchison, personal letter to the author.

Index

Page references in italics indicate illustrations.

About the Author

Susan Squier is Julia Gregg Brill Professor of Women's Studies and English at the Pennsylvania State University, University Park. She is the author of *Virginia Woolf and London: The Sexual Politics of the City*, editor of *Women Writers and the City: Essays in Feminist Literary Criticism*, and co-editor of *Arms and the Woman: War, Gender, and Literary Representation*.